The Master of Disguise ■ *My Secret Life in the CIA*

The MASTER of DISGUISE

My Secret Life in the CIA

Antonio J. Mendez

with Malcolm McConnell

Perennial
An Imprint of HarperCollinsPublishers

From *The Art of War* by Sun Tzu, translated by Samuel B. Griffith. Translation copyright © 1963 by Oxford University Press, Inc. Used by permission of Oxford University Press, Inc.

A hardcover edition of this book was published in 1999 by William Morrow.

THE MASTER OF DISGUISE. Copyright © 1999 by Antonio J. Mendez. All rights reserved. Printed in the United States of America. No part of this book may be used or reproduced in any manner whatsoever without written permission except in the case of brief quotations embodied in critical articles and reviews. For information address HarperCollins Publishers Inc., 10 East 53rd Street, New York, NY 10022.

HarperCollins books may be purchased for educational, business, or sales promotional use. For information please write: Special Markets Department, HarperCollins Publishers Inc., 10 East 53rd Street, New York, NY 10022.

First Perennial edition published 2000.

Designed by Carla Bolte

The Library of Congress has catalogued the hardcover edition as follows:
Mendez, Antonio J.
    The master of disguise : my secret life in the CIA / by Antonio J.
Mendez, with Malcolm McConnell.—1st ed.
        p. cm.
    ISBN 0-688-16302-5
    1. Mendez, Antonio J. 2. Intelligence officers—United States—
Biography. 3. United States—Central Intelligence Agency.
I. McConnell, Malcolm. II. Title.
JK468.I6M46   1999
327.1273'092—dc21   99-23066
[B]   CIP

ISBN 0-06-095791-3 (PBK.)

    04 05 06 FOLIO/ QW 10 9 8 7 6 5

■ *To all the Masters of the Game; some whom I met or engaged during the Cold War and others I came to learn about or know while writing this book.*

■ *To all the members of our families who also served the cause.*

Contents

Acknowledgments

I UNDERTOOK THE WRITING OF THIS BOOK WITHOUT FULLY REALIZ-
ing the complexity of such a project. Although many others have written
books about their careers in intelligence over the years and several have
done so since the end of the Cold War, none could have been so blessed
with encouragement and help from colleagues, friends, and family. Also
I received excellent assistance and advice from many highly qualified
and understanding people as this project unfolded.

My wife, Jonna, who was Chief of Disguise more recently and
worked at the CIA for twenty-seven years, was fully engaged in this project. I penned the first lines in November 1997 and the last change to the
manuscript was made on the fifth and final draft a year and a half later.
Her writing and editorial advice greatly enhanced the process and final
product and her creative judgment and political sense helped ease my
way more than once.

My collaborator, Malcolm McConnell, proved he has infinite patience. He first called me about *Reader's Digest*'s interest in doing an
article about my rescue of the six U.S. diplomats from Iran, and mentioned he had always wanted to write a book about CIA successes during
the Cold War. "So do I," I answered, and so we did. His wife and able
partner, Carol, was also a joy to work with and proved a marvelous cook
and hostess as well.

My agent, Andrew Wylie, and his assistant, Jeffrey Posternak, have earned a well-deserved reputation as the top guns of the literary trade. It is no wonder their legions of well-known clients trust them to run interference.

My editor, Betty Kelly, and her assistant, Alice Lee, have made this project first among many at William Morrow. Their thoughtful review and deft editorial changes, line by line, have added infinite value, well beyond what one might expect in the hectic world of publishing. They also developed and spread an interest and excitement for the book throughout their organization that bodes well for the further success of the project.

Thanks also to my friends and associates, formally or currently in the CIA, who read and made corrections on the various chapters. Everyone in this group was there in the midst of the Cold War with me. They each appear as one of the major players in their respective stories. All have been given pseudonyms as a matter of courtesy or good security practices. Some are still serving in harm's way while others are in CIA's most senior positions. They all took time away from busy operational schedules to help me write this book because they believe it is important.

A special thanks to the reviewers of the final draft. Their comments and suggestions helped ensure an independent point of view, plus the historical accuracy and technical quality of the work. This group includes: H. Keith Melton, author of *The Ultimate Spy Book*, noted espionage expert, and military historian; A. Denis Clift, president of the Department of Defense, Joint Military Intelligence College, and a senior staff officer in the National Security Council in the White House during many years of the Cold War; John Hollister Hedley, retired chairman of the CIA's Publication Review Staff and senior career intelligence officer;

Gordon E. Smith, a scholar and professor of Russian studies who studied and did research in Moscow in the late 1960s and early 1970s; and Catherine Eberwein, a senior staffer on the U.S. House of Representative's Permanent Select Committee on Intelligence and an expert on counterintelligence and counterterrorism.

Finally, my regards to the chairman of the CIA's Publication Review Board, Scott Koch, and his staff, who led me through the wilderness of the review process. Also my regards to the members of the board who have a tough job and little time to do it. By federal statute they have only thirty days to complete their review of sensitive material. The month they reviewed my final manuscript, they completed their review on over four thousand pages of material.

The board obliges me to include a disclaimer stating that "CIA's Publication Review Board has reviewed the manuscript for this book to assist the author in eliminating classified information, and poses no security objection to its publication. This review, however, should not be construed as an official release of information, confirmation of its accuracy, or an endorsement of the author's views."

The entire experience of preparing this collection of Cold War tales, while often harrowing, in the final analysis turned out to be as smooth as silk; not unlike my twenty-five-year career in the CIA.

Antonio J. Mendez

Preface

I DECIDED TO WRITE THIS MEMOIR IN SEPTEMBER 1997, WHEN THE Central Intelligence Agency publicly celebrated its fiftieth anniversary. During three of the Agency's five decades, which spanned the Cold War years, I served as a professional intelligence officer, creating and deploying many of the most innovative techniques of the espionage trade.

My purpose in writing this book, however, is not to bring credit to myself. I have already received ample recognition in the intelligence community. Vanity is not at stake in this project. Rather, I want this book to describe as accurately as memory permits a few of the operations my colleagues and I conducted. The reader can judge for himself the quality of our service in the cause of freedom.

Some of those we worked with are no longer alive. Others prefer to celebrate their achievements privately. Others are still actively engaged and must remain in the shadows. I have changed certain details of their identities so that they can remain anonymous. But, willing to err on the side of openness, I chose the potential risks of telling our story. I trust that doing so will also serve the cause of freedom.

Almost since its inception, the American intelligence effort has been either vilified by the world's news media—sometimes as part of Soviet disinformation operations—or romanticized by spy novelists with only vague notions of the nature of espionage operations. Yet for more than fifty years, Americans have been asked to support—both morally and financially—a large and active intelligence effort of which they have had little concrete knowledge. Several Directors of Central Intelligence and many of my colleagues have concluded that it is time to share more details of the earnest endeavor we made in the name of the American people. I agree.

I realize that my decision alarms certain intelligence professionals who see no need to breach the principles of silent service that I and others instilled in them during their training. But those who know me best will realize that I would never knowingly betray a trust or reveal a secret that would jeopardize a comrade, a source, or my country's interests.

Secrecy, of course, is the lifeblood of espionage. I am not a reckless renegade intent on exploiting clandestine operational methods to promote a book, nor do I feel it necessary to apologize for the U.S. government or the CIA's past errors or excesses. In telling my story, I intend to help redefine the CIA's traditionally stringent disclosure position. Although I will reveal much of what I know to be the Agency's ongoing contribution to preserving the world's peace and democracy, I intend to be consistent with sound security practices.

I must also adhere to the spirit of the law. Anything I write or say for public consumption is subject to scrutiny by the CIA's Publications Review Board. Further, I must consider the needs of intelligence professionals who continue to uphold the integrity of the service.

Publication review does not mean censorship. I have the same First Amendment right as any American citizen to express my opinion, positive or negative, about declassified details of our business. The review process is designed to determine whether any present or former Agency officers have violated the trust placed in them. We must all protect the appropriately classified aspects of the procedures and individuals used to collect intelligence vital to our country's security. The secrecy agreement we signed represents this contract of trust. We all signed it willingly, fully realizing that we would never be able to divulge certain details of our profession.

But attendant to this lofty obligation is a more practical concern, first expressed by veteran American intelligence officer Sherman Kent in 1955. He was proud of his leadership position but ambivalent as a scholar. Writing in the first issue of *Studies in Intelligence*, then a classified Agency in-house publication, Kent noted:

> Our profession, like older ones, has its own rigid entrance requirements and, like others, offers areas of general competence and areas of intense specialization. People work at it until they are numb because they love it, and because the rewards are the rewards of professional accomplishment.
>
> Intelligence today is not merely a profession, but like most professions it has taken on the aspects of a discipline: It has a recognized methodology; it has developed a vocabulary; it has developed a body of theory and doctrine; it has elaborate and refined techniques. It now has a large professional following. What it lacks is a literature. From my point of view this is a matter of greatest importance.

And where is that literature? In 1998, forty years later, CIA director George J. Tenet announced that because of "current budgetary limitations," plans to declassify records on significant Cold War covert actions conducted from the 1940s through the 1960s would have to be shelved indefinitely. His announcement caused both academics and the CIA's history staff to express their deep disappointment. Tenet, however, conceded that the CIA still has "responsibility to the American people, and to history, to account for our actions and the quality of our work." He added that the public needs the Agency's histories "to judge for itself the contribution made by the Intelligence Community to the successful conduct of the Cold War."

This sentiment is central to my decision to write a memoir.

Many intelligence officers did not live to see the end of the Cold War. The efforts of those who fought in the shadow world of the espionage wars, from the torrid backwaters of the Congo to the spy capitals of Vienna, Moscow, and Berlin, were never acknowledged in a parade or public memorial.

There are a few dozen stars carved in the white marble wall in the lobby of the Central Intelligence Agency's headquarters that serve as mute tribute to staff officers who died in the line of duty. But like the countless documents awaiting declassification, this stone memorial gives the American public no opportunity to celebrate the sacrifice these men and women made for freedom.

Perhaps this book will help honor the memory of their service.

1 ■ A Letter Slipped in the Door

Delicate indeed, truly delicate. There is no place where
espionage is not used.

    —Sun Tzu

The Blue Ridge Mountains, Maryland, August 21, 1997 ■ The anxious
memories returned to haunt me that summer night, keeping me from
sleep once more . . .

    It is past midnight near the time of the monsoon. I wait tensely on
the concrete observation deck of the sweltering airport terminal, peering
down at the tarmac through a thickening haze. The TWA flight from
Bangkok is already two hours late. I have watched Swissair arrive from
Riyadh, Lufthansa from Bangkok. An Aeroflot IL-62 arrives from Tashkent and lumbers up to the gate directly below.

    My pulse suddenly surges. The appearance of the Aeroflot is an om-
inous sign. The operations plan called for the subject and his CIA escort

to have left on the continuation of the delayed TWA flight at least an hour ago, for a very good reason. We wanted them out of here before the Aeroflot landed, with its inevitable ground retinue of KGB gumshoes.

The subject is a KGB defector who simply walked into our Station ten days earlier. Now, waiting down in the steamy, crowded departure hall, will he panic and run when he hears the Soviet flight announced?

I glance over the mildewed cement barrier. All the gates are full, but there is no American plane. Then, out of the gloom, the TWA Boeing 707 materializes. It lands, taxis down the runway, and finally stops at the far end of the poorly lit parking apron.

The haze thickens—"smit," the old Asian hands call it, ground-hugging "smoke from shit" from the millions of cow dung cooking fires burning in villages across the subcontinent. I squint, but the TWA plane is hard to distinguish. I wait.

The disembarking TWA passengers grope their way through the murk and stumble into the terminal, where the humidity and stench of clogged W.C.s will certainly overpower the smit.

I cannot leave the platform. My task is to confirm that our subject and his escort officer "Jacob," my partner in this operation, safely board the continuation of the TWA flight. But in this miasma, how can I see whether they reach the plane? If I don't catch sight of them coming out of the terminal with the other passengers booked for the same flight, it could mean they have run into trouble at passport control. That is where the alias documents and disguise I've helped create will be tested.

Passengers emerge from the terminal, headed for the TWA plane, but I still don't see the subject and his escort. Is it possible that they have already bolted to the two getaway cars sitting at the dark end of the parking lot with their engines running?

Whatever the outcome of the exfiltration operation, I have to pass a signal from the phone booth at the bottom of the stairway. Tonight, we will use an open code with an ostensible wrong number. Is Suzy there? (They made it.) May I speak to George? (Something went wrong.) The rest of the plan will unfold based on which of these two things happens . . .

Finally, I sleep, but I have no rest. Even in my dream, my mind cannot let go of the scene at the airport. I find myself descending the stairs with their chipped paint and wedging myself into the oven of the phone booth. I lift the receiver of the clumsy red Bakelite phone, put a brown coin in the slot, strike the cradle bar and release it. No dial tone. No coin drop. Damned colonial phone, a legacy of British rule that probably hasn't been maintained since the King folded the Union Jack.

Again I jiggle the cradle. The fat copper disk drops into the coin return slot. I jam the coin back in. A hiss, a click, a weak dial tone. Receiver held between ear and shoulder, I dial quickly, scanning the number scrawled on the hotel matchbook in my other hand. Clicks and pops, finally a coherent double whir. The phone is ringing at the other end. I press the receiver tightly against my ear. Four rings . . . five . . . *Pick it up, Raymond*. I slam the phone down after ten rings.

Why doesn't he answer? I look at my watch: 3:07, an hour past my scheduled call time. I know he's still at the safe house. They're expecting me to pass the signal. I suck in a deep breath of humid air and release it slowly to ease the tight band across my shoulders and the drumming in my ears. I *have* to call. I insert another fat copper coin and dial. A pause. A click . . . the coin drops through again. The phone is dead.

BLINKING AWAY SLEEP, I open my eyes to the wispy dawn spreading over South Mountain. The August sun brushes the treetops behind our garden teahouse. I blink again. Is that smit swirling among the azaleas? No, only mist.

The intensity of the dream dissolves slowly. I'm not in South Asia on an operation twenty-seven years in the past, but in the master bedroom of our house in the Blue Ridge Mountains. Still, as the cardinals start to sing, I am gripped by an anxious lethargy, the helplessness of the dream. Unable to return to sleep, I watch the colors in the garden change with the sunrise and quietly reflect on my life.

I've considered myself an artist since childhood. For a long time, I also saw myself as a competent spy. Since 1990, when I retired after a twenty-five-year espionage career in the CIA, I have once again been painting full time.

During these seven years of normal life, the recurrent dreams of the world I inhabited for so long have only slowly subsided. But one day, a totally unexpected event occurred that unleashed an avalanche of long-suppressed memories.

I LIVE, WORK, and show my paintings on forty acres in the Blue Ridge Mountains of western Maryland. Our post-and-beam house, and the surrounding studios on this lushly wooded property create a harmonious atmosphere similar to that found on a New England farm. A hundred feet down the grassy slope from the house stands a two-story studio, a red saltbox carriage house, and several sheds. Dominating the studio and stretching back toward the house is a large enclosed pavilion with a third-story tower perched in the center of the roof. The pavilion's second floor is a main exhibit area above a large office at ground level. My writing studio is now in the tower. This complex of wooden buildings is my personal work in progress, built by the hands of family, friends, and myself over the years since 1974.

Surrounding the studios and house are terraced gardens that my wife, Jonna, claimed as her personal domain when we married in 1991. Maple and oak trees cover most of our rolling property, on the base of a Blue Ridge summit west of South Mountain.

Late on Thursday, August 21, 1997, Jonna and I drove up the winding gravel track with our four-year-old, Jesse, asleep in the backseat of the red Pathfinder. From the garage bay beneath the studio, we passed the door of the office. A white envelope had been slipped into the screen door.

"What's that?" Jonna asked.

I got out and retrieved the envelope. "FedEx letter," I replied.

While Jonna tucked Jesse into bed, I examined the contents, a single-page letter on heavy bond stationery bearing an official letterhead:

THE DIRECTOR OF CENTRAL INTELLIGENCE

WASHINGTON D.C. 20505

The letter was addressed to me and signed by George J. Tenet, the newly appointed and just confirmed Director of the CIA. Its purpose was to inform me that I had been selected by my peers as a "CIA Trailblazer."

The Agency had established the Trailblazer Award as part of its fiftieth anniversary celebration. Fifty Trailblazers or their survivors would receive commemorative medallions during a closed ceremony at CIA Headquarters in Langley, Virginia, scheduled for the anniversary date, September 18, 1997. I would be among the "CIA officers who by their actions, example or initiative helped shape the history of the first half century of this Agency."

Tenet noted that veterans of the Office of Strategic Services (OSS), the Agency's World War II predecessor, as well as former CIA employees, had nominated three hundred candidates to be honored. A select panel had "worked very hard" to narrow the list down to the present fifty.

I read the letter again slowly, finding it hard to grasp that I was one of those selected.

Jonna poured herself a glass of cold water. "Anything interesting?" She assumed the FedEx was related to our art business.

I handed her Tenet's letter. "Take a look at this."

"Amazing," she whispered, shaking her head. Jonna herself had retired from the CIA in 1993 with twenty-seven years of service, so she recognized the significance of the award. Tens of thousands of people had worked for the CIA in the past fifty years, hundreds of them virtual legends in the intelligence community, but most unknown to the public.

Jonna read aloud from the letter, noting that I had been one of the people chosen out of all those "of any grade, in any field, and at any point in the CIA's history—who distinguished themselves as leaders, made a real difference in CIA's pursuit of its mission, and who served as a standard of excellence for others to follow."

I couldn't sleep that night, despite the cool breeze and the soothing chirp of crickets from the garden, so I got up and climbed the staircase to my small studio, to reread Tenet's letter.

On shelves around the desk were mementos of my CIA career. The dim light glinted off the tarnished silver of a Hmong necklace. I glanced at a framed picture of a boxy little Zhiguli sedan, driven along the Moscow embankment by a surveillance team from the KGB's Seventh Chief Directorate and reflected in the slush-spattered side mirror of an Embassy Ford. But one object stood out from the others. I reached up for a small case and removed the bronze Intelligence Star I'd been awarded for "courageous action" during a highly sensitive mission to Tehran at the height of the hostage crisis. It was a journey made in alias, using false documents—a hazardous and difficult assignment successfully accomplished, but never described in any unclassified publication.

Hefting the cool weight of the medal, I considered the closing comments of Director Tenet's letter. "Your achievements and those of the other forty-nine CIA Trailblazers probably will never be known in their fullness by the American people."

As I replaced the Intelligence Star in its velvet-lined case, I pondered the improbable sequence of events that had led me to this time and place. In some ways, I realized, I had been destined from childhood for a career in the shadow world of espionage.

I WAS BORN in 1940 in Eureka, an old mining town snuggled into the Diamond Mountains in central Nevada. As Route 50—"the loneliest road in America"—entered Eureka, it passed through a gap in a black wall of slag, the detritus of the enormous tonnage of silver and lead ore smelted earlier this century.

When the World War II boom hit Nevada, my dad, John G. Mendez, was hired at the nearby Kimberly copper mine. He was only twenty-three when he was crushed between two ore cars deep in a mine shaft. He lingered three days, then died on October 24, 1943, three weeks before my third birthday and the day after Mom's twenty-fourth. He left behind a young widow, four children, and a token insurance settlement. After the accident, we moved in with Mom's mother, Ina Bell, in Eureka.

It was in Grandma's old frame house that I learned of the family's pioneering history. My great-grandfather, Cristoforo Giuseppe "J.C." Tognoni, one of the legends of Nevada's gold bonanza earlier this century, had been born into a big family in the mountain town of Villa di

Chiavenna in the northern Italian region of Lombardi. J.C.'s father died in 1872, and the boy struggled to help his family survive before immigrating to America at age fifteen.

Somehow, he reached the United States and traveled west to Nevada to join two of his brothers. J.C. already possessed a skill highly prized in mining towns: In Italy he'd begun learning the secrets of the *carbonari*, who transformed wood into the high-grade charcoal needed to fire smelters. It was working in the mountains as a young charcoal burner for pennies a day that J.C. gained his intimate knowledge of the land forms and rock formations of Nevada. Still a teenager, J.C. headed off to seek his fortune prospecting in the Comstock Range.

A year later, he married Jesse Myrtle, a twenty-seven-year-old widow who worked as a cook on the mule-train line between Eureka and Tonopah. For the next fifteen years, the couple struggled, with J.C. working a succession of hard-rock mining, ranching, and freight-hauling jobs to stake his next prospecting expedition.

In May 1903, J.C. rode a horseback circuit to the top of a volcanic extrusion called Vindicator Mountain and studied the jumbled landscape below. After twenty-five years in Nevada, he could identify not only promising rock face, but also see from the narrowing of the washes where water might be found to work any claim. In quick succession, J.C. registered claims on a series of sites near a settlement known as Goldfield. These claims were among the richest gold strikes anywhere in the West. Within two years, J.C. became one of the wealthiest men in the state.

In 1916, my grandparents, Joseph R. Tognoni and Ina Bell Cates, eloped in Goldfield when she was just a teenager and moved to the Tognoni family's ranch. Although the cattle operation was making money, old J.C. was typically restless—"stubborn as hell," as Grandma always told us.

When he had sold some of his Goldfield claims in 1904, he had turned his prospector's eye to a likely spot at Black Rock Summit in the Pancake Range. Almost sixty, he pulled a horse trailer behind his Dodge Brothers truck until the dirt track petered out, then rode higher into the mountains. Beneath an eroded volcanic lip, he found a rich vein of reddish "ruby" ore and promptly named the claim Silverton. J.C. was convinced that he could make this remote mine pay because the silver vein appeared to run thick and deep through the volcanic ridge. So he began investing. But Silverton became the opposite of a mother lode: J.C. put hundreds of thousands of dollars into the claim, but did not take out a penny.

The banks eventually gave him an ultimatum: He could choose between foreclosure on the ranch or on the mine. J.C. chose to save the mine. He died at sixty-seven on August 9, 1932. No one ever did figure out how to make Silverton pay off. The old man was buried beneath that stark desert ridge, in a private cemetery where many of his descendants also now lie.

Eventually, my grandfather, J.R., took his family north to live in the old home in Eureka, but the stress of their financial collapse was too much for his heart, already weakened by childhood rheumatic fever. He died in February 1936, leaving Grandma with four children.

There wasn't much welfare in those days, and county assistance was especially slim in Nevada mining towns. So she and her kids learned to live by their wits and hard work in order to survive. Grandma had never driven her husband's '28 Dodge stakebody truck when he was alive, but that tough old vehicle soon became the family's principal source of money. Grandma bid on a "Lone Star" route, delivering mail to outlying mines and ranches, over washboard roads that switchbacked up stony mountainsides and crossed wide alkali flats. Her route led through some of the loneliest terrain in the state, and she drove through some of the most severe weather in North America. I grew up with eyewitness accounts of Ina Bell burrowing under her truck stuck in a snowdrift and cinching on her wheel chains before a sheepherder or passing busload of miners could stop and lend a hand.

My mother relied on Grandma's example to help her through the shock, grief, and fear that followed my father's sudden death. Mom went to work as the editor of the county newspaper, the *Eureka Sentinel*. She and Grandma pooled their money so that Mom could save enough for a down payment on our own small house. Then she met Arch Richey, who was twenty-seven years older than she was. They married in 1945.

Richey had worked in mines all over the West and lived by a simple rule: Give the boss an honest day's work and have a good time at night. His idea of a good time was drinking boilermakers, swapping yarns, and gambling.

Then the war ended, and the bottom fell out of copper and zinc. We left Eureka in 1947, the entire family jammed into the family car, a 1930 Model A Ford pickup piled high with bedding and suitcases. My half-sister, Maureen, was a baby, and Mom was expecting her second daughter, and final child, with Arch. The only work Richey could find was cutting rock at a quarry outside of Sparks, a suburb of Reno. He planned to build a house on the side of a desert mountain and had made a fair start on the foundation. But we were forced to spend several months in a sun-faded old Army surplus pyramid tent, sleeping on canvas cots, before moving into the shell of the house. We had no toilet and used a boulder slide down from the building site as the latrine. Our only running water was the cold Truckee River, where we washed our clothes and took baths, and where I learned to swim in the fast currents.

Twenty-some years later, these tough conditions helped me adjust to even more austere living when I'd visit the CIA's Lima Site strongholds on sheer limestone monoliths above the rain forest north of Vientiane and on the Bolovens Plateau in southern Laos. Night and day, heavily armed Pathet Lao and North Vietnamese Army troops moved in the nearby lowlands along the Ho Chi Minh Trail, hidden beneath the forest canopy. The sites could be supplied only by helicopter or rugged little short takeoff and landing planes, so amenities like water were precious. Still, taking a sponge bath from a five-gallon Jerry can that had warmed all day in the tropical sun was a lot better than bathing in the Truckee River in the winter.

It was in the half-finished house in Sparks that I started to draw. Using a brown paper bag and a carpenter's pencil, I created a primitive cartoon strip of my family's trek from Eureka and the progress of the new home.

One Saturday after payday, Mom came back from town with small presents for the kids. "This is for you, Tony," she said, handing over a sketch pad and box of watercolors. "You have the qualities of an artist."

∎

AFTER LIVING ON that mountainside in Sparks, the drafty house above the tracks in Caliente did not seem like much of a hardship when we moved there in 1948. But there were plenty of nights between paydays when the single pot of beans Mom always kept on the stove and a few slices of Wonder bread had to stretch to feed us all.

On summer mornings, my brother John and I often headed up the canyons, lugging a black cast-iron frying pan, a potato, and an onion, sometimes an end of bacon if we were lucky. I knew where the rabbits watered, so we usually shot one and had it sizzling in the frying pan by noon. Then we'd doze under the scrub cedars through the hot afternoon before climbing the steep pink rhyolite cliffs to chisel out a gunnysack of rock-hard bat guano—which, strangely, was always laced with cactus needles—from caves I'd found my second summer in Caliente. High-quality bat droppings made excellent fertilizer, and we could usually peddle a gunny sack for a dollar to the better-off Mormon ladies across the tracks.

Word got out that we knew of secret caves, and some of the kids tried to tail us. I learned my first lessons in surveillance evasion by leading them into dead-end box canyons well away from our precious bat caves, then climbing out unseen through hidden, narrow chimneys.

John and I shared a *Las Vegas Sun* paper route and always asked our manager for extra copies when we filled our sacks each morning. Every afternoon we'd peddle these papers in bars along Main Street. We turned over our regular pay to Mom; anything we made selling the extras was movie money for all the kids in the family. Watching the Bob Steele and Superman serials at the Rex, I felt an early fascination with the magic of deception: The kids around me in the splintered old movie-house seats actually believed the characters were real. I knew they were actors, and I wanted desperately to discover what makeup techniques, props, and camera work had transformed them into celluloid heroes and villains.

Sometimes when we had extra papers, we'd hike out to Cherry Hill and sell them to the girls, who often had curlers in their hair and were just finishing their late-afternoon breakfast before a long night's work. I don't think John knew what a whorehouse was, but I was fascinated by the tangible sense of the clandestine that we would encounter by stalking over the ridges and surprising the married ranchers and store owners. Startled, they would jump back in their trucks and get away, terrified of being recognized by two scrawny little kids from town.

Their embarrassed expressions stayed with me for decades. I saw this again in Vientiane on the faces of government officials and diplomats leaving the White Rose or stumbling down the steps of Lulu's Rendezvous des Amis on the Mekong embankment. I would also encounter this caught-in-the-act resignation among Soviet bloc diplomats in Bangkok. When I was a kid, the knowledge that a "proper" merchant indulged in occasional pleasure with the girls on Cherry Hill was merely amusing. In Southeast Asia, spotting the second secretary of a Soviet embassy or an Indian military attaché at a whorehouse was potentially valuable information, so CIA stations organized regular "pole patrols" to conduct this unsavory but necessary form of surveillance.

The paper route provided another crucial skill that served me well in my later profession: the ability to deceive with plausible denial. When John and I were stuck with unsold papers from the previous day, we met the Union Pacific streamliner from Salt Lake to Las Vegas that stopped in Caliente for just nine minutes each morning. John would take the Pullmans, and I'd hit the restaurant car, calling out "Papers . . . get your morning paper!" The passengers were usually so sleepy that they handed over their dime without checking the paper's date. The fact that they'd just paid for yesterday's news wouldn't hit them until the train was well south of town. And it was unlikely we'd ever see them again.

But before beginning this little operation, we worked out a contingency plan. Growing up in remote mining towns without television, I'd spent hours studying magic tricks in my uncles' battered old copy of *The Boy Mechanic,* a book published in 1905 that taught everything from building gliders to making bicycle-powered washing machines and had several chapters devoted to sleight-of-hand and parlor-game illusions. I realized that tucking a couple of the correct morning's papers behind the stacks under our arms was just a variant of the old transfer-the-sugar-cube trick.

One day I sold a paper to a stocky professional gambler in a herringbone suit who was too busy chomping down ham and eggs to even look up as I took his dime. But as I turned away, he reached out and grabbed my shirt.

"What the hell is this?" he demanded, waving the *Sun* in my face.

I looked into his hard eyes, then noticed the ruptured duck pin in his lapel. He was a veteran, probably a former officer from the look of him, a guy who'd seen every yardbird swindle and heard every stockade lawyer's alibi in the army. I knew he'd have a sharp eye for sleight-of-hand if he'd spent much time at the gaming tables of Las Vegas and Reno.

"Oh, sorry, sir," I said, blinking at the paper's date. "They must have been upside down in my bag. I have to take the old ones back to the station." I turned over the stack, slipped him that morning's edition and retrieved the offending copy.

Even though there were several potential customers seated at tables down the line, I opted for a tactical withdrawal.

Chalk up another lesson that later served me well: Keep your options open; always have a fallback when you're working in hostile territory.

WE MOVED TO Denver in 1954, the year I finished eighth grade. At Englewood High School, I fell in with a group of boys who shared my interest in art. This "greaser" crowd liked to draw, but we were also fascinated with anything mechanical. We worshiped Marlon Brando and his gang in *The Wild One* and, of course, James Dean in *Rebel With-*

out a Cause. We drank beer, organized illegal drag races, harassed the local cops and our high school teachers, and had the occasional rumble with kids from outside neighborhoods.

While doing all right academically, I couldn't accept the smug conventions of Englewood High's reigning clique, the "sosh." During my junior year, 1957, the student government arbitrarily decided the Christmas dance would be couples only. This meant that guys like my buddy Doug and myself, who liked to go to dances to meet girls, would be excluded.

"Looks like we're out of luck," Doug said, applying a coat of wax to his 1932 V-8 Ford deuce streetrod in the old garage behind his mother's place.

"The hell we are," I said. Ever since the announcement that morning in home room, a plan had been forming. "We can go as a couple and tear the place up."

Doug grinned, but then looked worried. "Who's going to be the girl?"

I tossed a coin . . . and lost. I had two weeks to become "Denise," Doug's date from out of town.

I suppose the Christmas dance caper was my first covert operation, because it involved analyzing hostile security, planning an elaborate "cover legend," inventing a persona, and developing a disguise that would pass the closest scrutiny. Once Doug registered Denise as his official date from another school, we got down to the details of building a foolproof disguise. It had been announced that the regular school teacher chaperons would be supported that year by city police under strict instructions to maintain order and admit only couples with tickets. Their vigilance would discourage the leather jacket crowd from crashing the party.

I approached with dead seriousness the challenge of becoming a convincing teenage girl. Enlisting Mom's support was not hard; she relished the idea of putting something over on the "popular" kids. Mom borrowed a dress from a friend who was about my size, constructed a padded bra, and even plucked my eyebrows the afternoon of the dance. Doug and I had already picked out a long brunette wig from a costume shop in Denver. That evening, I shaved my legs and practiced my backward dance step, wearing the borrowed pumps for the first time and coached by my four giggling sisters.

As we moved down the line of couples toward the door of the school cafeteria, I could feel Doug's arm tighten inside his powder-blue sportscoat. There was a big, serious-looking police sergeant standing beside the ticket taker. It was not too late to back out, but I struck out my chest and strutted ahead.

"This is going to be so much fun," I murmured in falsetto.

"Shut up!" Doug moved ahead.

Out on the crowded dance floor, my anxiety dissolved and I began to enjoy the power of our deception. This was fun. Dancing nearby was one of the worst of the stuffed shirts, a senior named Dave, who'd been voted the school's "most studious." With me maneuvering backward, we slid in beside Dave and his date and I began to wink at him. Dave was so flustered that he stepped on his girl's white pumps. After two more dances, Doug and I pulled up alongside Dave again and unleashed the surprise we had been saving up for him.

"Bastard!" I shrieked, startling the entire dance floor.

"Bitch!" Doug snarled back. He threw a solid punch into my ample right breast. Girls around us screamed. I snatched off the brunette wig and threw it at Doug's face. He ducked and a girl behind him was hit by the flying tresses. For a couple of minutes we traded movie punches and wrestled, my crinoline skirts whirling. The desired effect was achieved. The dance's Goody Two-shoes snobbery had splintered into chaos.

As the cops and chaperons shouldered their way onto the dance floor, I snatched up the wig, and Doug and I sprinted for the emergency exit.

Although we expected to be called into the principal's office, we actually escaped punishment. Everybody in authority, including the cops, had found our little act outrageously funny. For me, the most amazing aspect of the episode was that a good cover story, supported by a clever disguise, actually could transform one person into another. But, as I would later teach hundreds of CIA case officers, disguise was more than a matter of putting on a dress and renting a wig. You had to *live* the deception. Every aspect of the altered persona—the walk, the voice, the posture, and the mannerisms was essential. They all combined to make a convincing whole.

AFTER A PLEASANT year at the University of Colorado in Boulder, I spent the summer working fifty-four hours a week as a plumber's helper, digging ditches and chopping through concrete with a jackhammer.

There was a building boom under way in Denver. The Cold War was heating up again after the cease-fire in Korea. Martin Marietta's aerospace division had just landed air force contracts to build Titan I and II intercontinental ballistic missiles (ICBMs). Engineers and skilled craftsmen were pouring into Denver from around the country.

One night in June at a swimming pool, I met a pretty girl with auburn hair named Karen Smith. A junior at Englewood High, she was about to turn sixteen. We dated all summer, and when I went back to Boulder that fall, I was sure I'd met my future wife.

But college wasn't the same. Even though I'd earned $1.85 an hour with plenty of overtime all summer, Mom had borrowed most of my savings because Arch was unemployed and still running up hefty bar tabs. The news from home that winter got worse. Money was so tight that there were times the utilities were shut off and the kids had to stay in bed, bundled under blankets, instead of going to school. I decided to take a semester off to help them out.

This time I wasn't as willing to overlook Arch Richey's problems as I had once been. We had some serious fights, and I was soon living in my own apartment and working again as a plumber. This gave me more time alone with Karen. We planned to get married when she graduated from high school. But, hormones and emotions being what they are at that age, a year's wait seemed impossibly long. In 1960, unmarried couples didn't just move in together, so we did the only thing we could: We eloped that May, on Friday the thirteenth.

■

FIVE YEARS LATER, Karen and I had three children, Amanda, Toby, and Ian. I'd gone from plumber's helper to a partnership in an art and design fabrication business, then landed a job on the night shift at Martin Marietta as an artist/illustrator in the proposal department.

The work at the Martin Marietta plant was neither exciting nor challenging. An engineering team would cobble together a proposal on a subcomponent for a military satellite, missile, or perhaps even a NASA spacecraft, and we'd have to transform their drawings into finished schematics and charts and provide an eye-catching brochure cover, often working all night.

Defense work was a roller coaster. I got advance word of my layoff a month before the hammer fell in 1963. Still on the Martin payroll, I qualified for a mortgage. Karen and I bought a house in the suburb of

Littleton that had enough room for a studio, shop, and gallery. I planned to make my way in the world as a painter and run the art and design business as sole owner.

But I soon found out that trying to live on a fifty-dollar-a-week unemployment check while hoping to sell landscapes to other unemployed aerospace workers was not a winning proposition. We had a couple of rough years during which I hustled for any kind of fabrication or store-decoration job, and even worked as a process-server for my lawyer uncle, Robert Tognoni. For a while, it looked as if my own kids were destined to live through the financial instability that had plagued my own childhood.

Then I got called back to Martin, this time as a tool designer working with electronic engineers on the huge new Titan IIIC booster. We were designing the missile's electronic control modules, which had to withstand the stress of launch and the hostile environment of space. My job took me into vast sheds where prototypes of the missile lay in cradles wider and longer than the huge Ute dump trucks in the Nevada mines or the gondola cars that used to rattle through Caliente. I had to crawl around the "bird," making exact measurements for my life-size drawings of the wiring harnesses and junction boxes.

The first time I actually touched the smooth titanium skin of the massive rocket, I felt a twinge of excitement mixed with dread. This missile might actually be fired in wartime, and I knew the possibility was not so remote.

The Cold War was no longer a nagging geopolitical dispute; it was a smoldering potential holocaust that could easily destroy human civilization, not simply individual nations. The wars in Korea and French Indochina had ended in bloody stalemates; Europe was divided by the

Iron Curtain. None of us would ever forget the tense October days of the 1962 Cuban missile standoff. Now an arms race of unprecedented magnitude had begun.

In the years since I'd witnessed my first atomic bomb tests in Nevada, both the United States and the Soviet Union had developed hydrogen bombs. Here in the plant, the blunt gray nose cones were mock-ups, but we all realized the Titan IIIC was meant to deliver halfway around the world a thermonuclear warhead yielding the equivalent of millions of tons of TNT. Just as the atomic bomb had made World War II blockbusters seem like firecrackers, the H-bomb, driven by the fusion process that fueled the sun, had turned fission weapons into toys.

IN FEBRUARY 1965, George Adams, a friend from my first job at Martin, dropped by my studio. He was still out of work and had answered a want ad in the *Denver Post* asking for "Artists to Work Overseas—U.S. Navy Civilians." George's résumé had been rejected.

"I had some juvenile police trouble," he explained.

I studied the ad. Working for the Navy "overseas" would probably be more exciting than drawing plans for wiring harnesses on the Titan missile. I sent in my résumé to the Salt Lake postal box listed in the ad. Ten days later, a man called and asked that I appear for an interview at a motel on West Colfax Avenue in Lakewood. "Funny," I told Karen, "I wonder why somebody from the Navy didn't want to meet downtown at the federal building."

Richard Ryman offered me a chair at the faux-wood-laminated motel table. A shaded bulb hung low above us, illuminating my chair, while Ryman remained in the shadows. The scene reminded me of a Sam Spade movie. Ryman was in his forties, tall and rangy with blond hair going to gray. He wore a snap-brim hat and no suit coat. His dark tie was loosened to expose a prominent Adam's apple at his unbuttoned shirt collar.

Ryman placed an open bottle of Jim Beam and two motel glasses on the table. "Care for a drink?" His tone was casual.

This was unusual behavior for someone recruiting a civilian Navy artist. But I tried to be friendly. "Sure," I said, holding out a glass.

As we sipped our bourbon straight, he slid a thick black ring binder into the light. "What kind of art work do you do?"

"I've got some samples with me." I bent to open my two old leather portfolios. For the next twenty minutes, Ryman sat silently, nursing nips of bourbon as I rattled on, showing prints of portraits, landscapes, specialty design projects, draftsman and architectural drawings, and the covers of Martin proposal brochures. Finally, I ran out of samples and sat down.

Ryman knocked back the last of his drink and opened the ring binder to a page he had marked. "I really don't know what kind of artist they're looking for," he said, gazing at me reflectively from beneath the brim of his hat. "I sent in several applications, but none of them seemed to be right. Here, you read this."

He swiveled the binder across the table. The bold red typeface leaped out at me from the top and bottom of the page: TOP SECRET—NO FORN DISSEM. The only classified documents I'd ever seen had been at Martin, and I knew they weren't supposed to leave a secure facility.

Then my eye moved down the page. I was staring at a recruitment guide prepared by the Central Intelligence Agency, Technical Services Division.

Half an hour later, I left the meeting with Ryman, a sixteen-page application form tucked inside my portfolio. At the door, Ryman touched my arm. "If you do apply, you *will* be polygraphed and subjected to a thorough background check to make certain you're trust-worthy and not an agent of a hostile organization." His tone softened slightly. "But remember, Tony, we're not looking for paragons. We don't give a shit if you patted someone on the ass once or twice. You just have to level with us."

Loading my portfolios into the car, I looked back at the motel room. The curtain was firmly closed, betraying no hint of the person inside.

2 ■ Border Crossings

Gentlemen do not read each other's mail.
> —Attributed to Secretary of State Henry L. Stimson,
> c. 1929

Washington, D.C., April 1965 ■ The bus from Baltimore Friendship Airport dropped me at the airline terminal on the corner of 12th and K streets on this bright spring afternoon. I paused for a moment on the curb, a pilgrim in his mid-twenties, trying to absorb what had been a memorable day. Not only had the American DC-8 flight from Denver been my first ride in a jet, it had also been the first time I'd flown. Now I was standing at the taxi rank, only blocks from the White House, my mother's worn leather valise at my feet, about to head across the Potomac to the CIA's Headquarters in Langley, Virginia. The sun was warm and trees were budding. *How nice it would be to stroll toward the mon-*

uments on the Mall, I thought. Until that moment, they had only existed in my mind as images from magazines and television.

But I felt a sense of duty. The letter I'd received inviting me to an interview had been clear: I was to report to CIA Headquarters at nine A.M. on Tuesday, April 20. Now I wanted to find a room in Langley and try to get a good night's sleep, *if* my excitement permitted.

"Langley, Virginia," I told the black cab driver, who wore a colorful, geometrically patterned shirt with billowing sleeves. It was the first time I had ever encountered an African.

"Where in Langley, mister?" He removed his toothpick and gave me a dubious glance in the rearview mirror.

"Any motel near the Central Intelligence Agency," I said.

"You say the CIA, mister?" The driver shook his head in mild amusement.

We cut onto Constitution Avenue and rolled along the Mall with the rush hour commuters streaming toward the Virginia suburbs. I couldn't help rubbernecking as we passed the Washington Monument, then the Lincoln Memorial, rising through the blossoming trees ahead. To me, this was *infinitely* more exciting than diagramming circuits at the Martin plant in Denver. I felt as if I was finally being lifted from the isolation I'd known as a kid, and the tough times I'd endured as a young artist, into an exotic world of adventure.

But as the taxi weaved among the commuters crawling on the George Washington Parkway on the other side of the Potomac, the mundane returned to quash my sense of adventure. "You got a name for that particular motel, mister?" the driver asked with a wry grin.

"Anywhere near the CIA," I replied, trying not to sound like a naive tourist.

He cut up the steep road beside Chain Bridge and turned onto

Route 123. I assumed we were in Langley, but there was nothing but miles of newly budded oaks, maples, and sycamores. "Down there is the CIA," the driver said, proud of his knowledge. I could barely make out what appeared to be a high security gate and a white concrete-block guardhouse, half-hidden by tree trunks, about three hundred yards down a curving, two-lane road. A small sign at the shoulder read simply: VIRGINIA HIGHWAY DEPARTMENT. I was suddenly shaken by uncertainty. Had all this business—the mysterious ad, the bizarre interview with Ryman, and now this taxi ride to nowhere—been a hoax?

"Where's Langley?" I looked around anxiously. "Where's the town?"

"Don't know that there is one," the driver said, his smile widening with the knowledge that he, an immigrant, knew more than a native son. "We will have to have a look up here by McLean."

An hour later, we were still searching for a motel, but the driver, who revealed he was a "very good Ghanaian Christian man," had assured me I wouldn't be charged more than the standard fare from the District to McLean. Finally, in Falls Church, we spotted an old Victorian house with a white gingerbread porch and a small blue sign that read ROOMS FOR RENT. We were a good five miles from the entrance of CIA Headquarters. Could I get a cab in the morning, or would I have to hitchhike?

"The city bus to the CIA stops right here at the corner," the landlord said. I paid my Ghanaian driver a generous ten dollars for his trouble and unpacked the suitcase in my temporary base of operations, a second-floor bedroom shielded from the street in a quiet neighborhood—my gullible idea of the operational security this first visit to Agency Headquarters required.

But the next morning, I was again dragged back to the commonplace that lies beneath so much of the cloak-and-dagger sensationalism of the intelligence profession. The municipal bus was crowded with commuters. At the CIA stop, I got off with about a dozen people, some carrying brown-paper lunch bags, and trudged along toward the security hut. I noticed that nobody looked like James Bond. The shapely Miss Moneypenny was also nowhere to be found.

As we rounded the curve, however, I got my first look at the immense white limestone monolith of the Agency's new Headquarters, reputed to be the second-largest office building in the world after the Pentagon. I had no reason to doubt that this massive, seven-floor structure dominating a green campus, although invisible from the surrounding roads, was in fact 1,400,000 square feet. The clean sweep of the roof was broken by clumps of antennae. An enormous fiberglass igloo, which I guessed must have sheltered satellite dishes, stood to the right of the building. Suddenly, I forgot about the brown-baggers and thought about the cryptic messages streaming to and from this building. As I waited in line at the uniformed–security guard station, my sense of awe and excitement returned.

The people in line flashed blue laminated badges, and the guards waved them through. I had my interview letter open like an eager scholarship boy on the first day of school. "I have a nine o'clock appointment with personnel," I told the guard, handing over the page embossed with the CIA seal. He glanced at the letter, then directed me to the swooping glass-and-marble front portico. Walking past the mysterious igloo, which I soon discovered was actually the auditorium, I imagined agents punching digits into tiny burst transmitters, just as I'd read in Len Deighton novels. This mental Ping-Pong match between the exotic and the workaday existence of bureaucracy had only just begun.

Standing on the polished gray granite floor of the vaulted marble lobby, I gazed at the imposing seal, thirty feet in diameter. The words

THE MASTER OF DISGUISE 27

CENTRAL INTELLIGENCE AGENCY OF THE UNITED STATES OF AMERICA were inlaid in multicolored granite characters, surrounding a shield topped by the profile of a fierce eagle. The center of the shield was emblazoned with a sixteen-point compass rose, symbolizing the far corners of the world where the Agency operated. High on the marble lobby wall was the passage from John 8:32, . . . AND YE SHALL KNOW THE TRUTH AND THE TRUTH SHALL MAKE YE FREE . . . If the architect intended to impress people with this lobby, he had succeeded in my case.

"Your first appointment is at the West Out Building," the personnel officer told me, consulting a neatly typed schedule. He wore glasses with heavy black frames and kept a water glass filled with at least a dozen precisely sharpened pencils on his spotless desk blotter. Slender and balding, the man did not fit my image of a spymaster. "You'll have to catch the Bluebird at the stop out front," he said.

The Bluebird was an unmarked bus that swung by the Pentagon before crossing Memorial Bridge, making a couple of intermediate stops at anonymous buildings, then depositing me at 15th and Independence on the Mall. It was another bright spring day, and I was once more torn between my sense of obligation and my desire to soak up the sights. But I wasn't prepared for the ranks of dingy prefab "temporary" buildings dating from World War II that sprawled across this slope of the Mall and certainly never made it onto tourist postcards.

The West Out Building was standard red brick government from the nineteenth century, a gritty facade broken by rows of windows too narrow to have offered much relief to the bachelor clerks in celluloid collars sweating at their desks in the decades before air conditioning. From the sidewalk, there was no way to discern the building's current tenants; it could easily have been a branch of the nearby Bureau of Engraving and Printing. But in fact, it was the front office of the CIA's

Technical Services Division (TSD). Here I would have to pass my initial screening before being led across borders of secrecy into the concentric circles of security guarding the Agency's clandestine heart.

That morning's interview was with two gentlemen I will call "Phil" and "Franco." Phil was the chief of the Graphic Arts Reproduction Branch of TSD, and Franco headed the Art Department. These two men had the power to determine my future.

Phil held up the art samples I had sent him with my initial application, including some rather elegant oriental calligraphy. "Why were you interested in this type of thing?" Franco asked, keeping his voice neutral.

It was obvious they were intrigued with my samples, and I certainly didn't want to be deceptive. "I saw the recruitment guidelines during my Denver interview," I admitted. "So, I figured part of this job might involve, well, reproducing foreign writing."

Both men offered cautious smiles. Was I being resourceful or too clever? Maybe both were attributes the Agency desired, I silently hoped.

"That explains it," Franco said, stacking the pages neatly in a folder. "Not many artists send in Chinese grass writing or a watercolor of a Bulgarian postage stamp. We wondered if you were particularly insightful or some kind of a con man."

Franco was a large, jolly man with a few surviving wisps of hair. Now his smile was genuine. "It appears you're aware of what we do here."

"I read that you call it authentication and validation," I offered, referring to a book I'd borrowed from the Denver library about the CIA and the Technical Services Division.

Phil, a dour no-nonsense type, nodded, impatient to cut through the pleasantries. "You understand that you will be reproducing documents, then?" His cool eyes locked on mine. "That doesn't leave much creative latitude. Your lettering and line weight will be critiqued with a microscope."

In other words, a major part of my job would be duplicating official documents of foreign governments, ranging from ration books to letterheads to military identity cards. If I did that type of work on American soil with U.S. documents, I'd be in line for a ten- or twenty-year vacation in Leavenworth. I sensed that Phil was probing to see if I had the stomach for intelligence work in general; he was also testing my ability to surrender artistic freedom and succumb to the drudgery of providing the forged documents needed to sustain the Agency's clandestine operations. I remembered the oft-quoted phrase of former Secretary of State Henry L. Stimson, who was reputed to have said, "Gentlemen do not read each other's mail," when he effectively shut down American intelligence operations in 1929.

Things had changed. The Cold War was entering its twentieth year. Not only was I willing to read the mail of any Soviet gentleman I might encounter, I was eager to take a crack at forging some, if it furthered my country's interest. "I understand creative problem solving, which is what I think you're doing here," I told Phil. "The mechanical processes involved are incidental. The real creativity is in the planning and management."

That particular kernel of wisdom had come from one of the bullpen artists back at Martin Marietta, who had been studying for a management degree.

"Well, all right, then," Phil said, nodding to indicate he'd made his point. "I just don't want us to have any misunderstanding."

I knew I had passed the first hurdle.

The next appointment on my sheet was with a fellow named Glenn who headed a new TSD program bearing the acronym TOPS (Technical Operations Officer Generalist). He explained that the Agency's technical operations had always been a vital part of our espionage efforts, dating to the CIA's World War II predecessor, the Office of Strategic Services (OSS). But, at its inception in 1942, the "Oh, So Social" had in fact drawn a lot of its officers from the prep school–Ivy League network, while their sergeant radio operators, interpreters, and demolition men had often been patriotic blue-collar draftees. Since the Agency had originated in 1947 at the start of the Cold War, Glenn assured me, any sense of class difference between case officers and the TSD had disappeared.

I would soon learn more about the TOPS program, which assigned certain qualified TSD officers overseas to work side-by-side with their case officer colleagues. For example, a case officer might be skilled or lucky enough to recruit an agent willing to plant a listening device in the desk drawer of a sensitive department of a foreign government. Building the device and training the agent in its use were the TSD officer's jobs. TOPS officers could handle most urgent jobs, but they knew when to ask for help.

"I see you're working on spacecraft electronics, have practical experience in construction, and have even run your own fabrication and design business," he said, checking off items on my sixteen-page application. "There might be a TOPS assignment for you in a few years." I later realized that Glenn was assessing my potential for what was known as a more exciting "singleton" assignment overseas—the forward deployment of tech ops (technical operations) officers as TSD firemen. They were expected to fly in on short notice and provide the technical support that case officers might need in a fast-breaking operation, perhaps in a remote area.

That was good news because I didn't relish the idea of spending my entire career hunched over a drafting board, forging military train passes for the Deutsche Demokratische Republik. I also wanted to see such documents actually being used.

After lunch I met with Dr. Sydney Gottlieb, the Deputy Chief of TSD. As I entered his high-ceilinged office, I couldn't help thinking of the apochryphal Q in the James Bond sagas. A huge framed replica of Goya's *The Nude Maja* hung on the wall behind him; on his desk was a tiny plaque reading THINK BIG. Gottlieb looked more like a tanned, athletic college professor than someone who worked in a basement laboratory, dreaming up smoke-screen dispensers for gleaming sports cars, or cunning devices used to eavesdrop on telephone lines without leaving a physical trace. Little did I know . . .

Hardly stuffy, he sat with me in front of his desk with his shirt sleeves rolled up. I was conscious of his careful grooming and deliberately casual manner. "I like to meet with everyone who's being considered for employment with TSD," he said. After asking about my family, he looked at me thoughtfully and paused for a moment. "How would you feel about telling a lie for the good of your country?" he asked. There was a probing intensity to his question. I resisted the urge, natural in a young man with his foot in the professional world for the first time, to give a glib answer. "A lie can be a deceitful way to hurt people," I finally said. "Or it can be a necessary form of deception in time of war. I think we're engaged in a war, and both sides are using deception."

I sensed that my answer met his demanding criteria.

At the end of my interviews in the West Out Building, Manny Fontana, TSD's chief of personnel, wrote out a brief note on an Agency form. "Report back to Headquarters tomorrow morning for a polygraph and a full physical," he said. I suppressed the urge to grin, having survived the first row of hurdles. Manny added that if I successfully passed the physical and polygraph, I'd be offered a position in the TSD Art Department at the level of GS-9/1 for the princely salary of $8,672 per annum. But, he cautioned, the offer was contingent on a background investigation, which would take a few more months to complete.

It was mid-afternoon when I walked out of the old brick building and onto the sunny grass of the Mall. Now I could smile and hoot in delight—I was elated. The wider world of daring exploits that I had dreamed of as a kid those nights in Caliente, listening to the scratchy voices of London and Cape Town on my beat-up old radio, seemed to lie open at my feet. I spent the rest of the afternoon at the National Gallery of Art, awed to be in the presence of Rembrandt, Titian, and Picasso.

THE ROOM WAS small, quiet, and softly lit. I could have been in the empty waiting room of a heart specialist. Instead, I was sitting in a comfortable blue recliner in one of the Agency's polygraph suites on the second floor of Headquarters. I tried to relax and breathe normally, but I was conscious of the elastic respiration band across my chest and the wired pads clipped to my fingertips. Worst of all, the inflated blood pressure cuff clamping my left biceps was so painfully tight that my arm was going numb down to my fingertips.

"Is your name Antonio Joseph Mendez?" The middle-aged man sat at the small table close to the left arm of my chair, concentrating on the machine before him. A paper scrolled from a spool in the polygraph, its surface a network of smooth wave lines from the sensor scribes. The man's voice was soft and kind but carried a sense of purpose.

"Yes," I answered, thinking, *What a bizarre question*. Then I realized I could have been there under false pretense. What better way for a foreign intelligence service to penetrate the CIA than to have one of its own people assume the identity of an American who contrived to have himself hired? In addition, answering such a straightforward question was probably designed to establish a baseline of truthfulness, a hunch that was confirmed when the polygraph operator asked me to verify my address in Colorado and my place of employment.

In the same cordial tone, he shifted his approach. "Are you working for any organization hostile to the United States?"

"No."

"Have you ever told a lie?" His voice was completely devoid of confrontation.

Obviously I had. "Yes." The man didn't even look up from his chart.

"Have you ever divulged a confidence?"

Yes, who hasn't? But I answered, "No." After a moment's reflection, I changed my mind. "Yes."

"Have you ever divulged a confidence?" he repeated, ticking a corner of the scroll with a colored pencil.

"Yes." These last two questions, I saw, were also probably intended to assess how a lie might register. Only a saint could answer "no" with a straight face, and the way I replied probably gave him a good indication of how one of my true, positive responses registered on the sensors that day.

Over the next twenty years, I would be regularly "boxed," as were all employees of the CIA, and I would come to learn why polygraph evidence is not admissible in a court of law. The "box," in the hands of a skilled polygrapher, could at best detect patterns of deception, not consistently distinguish individual lies from truthful statements. But as a screening device for recruits, both American officers and foreign agents, a well-run polygraph exam probably remains the best technique available to unearth deceit, some notable failures notwithstanding.

". . . and what were you thinking when I asked if there was anything else you wanted to tell me?"

I chuckled, in spite of the painful grip of the blood pressure cuff. "I was thinking about the TV show *The Fugitive*. I'm not sure why, it just popped into my head."

Five minutes earlier, he had asked me if I'd ever stolen anything, and I'd admitted stealing a bale of hay from a barn when I'd been a kid. This, he proclaimed, was "of no consequence."

Now he switched to another tack. "Have you ever used any illegal or controlled substance?"

In 1965, where I came from, this could mean only one thing—marijuana. "No."

"Have you ever been involved in sexual activity that might be used against you for blackmail or coercion?"

"No." The chart purred softly off the machine.

"Criminal activity?"

"No." I thought of my minor scrapes with the law as a kid in Denver. I even remembered peddling day-old newspapers on the Union Pacific. But apparently, my answer had been satisfactory.

He bent over to remove the cuff and the galvanic skin-response pads from my fingertips. "Now," he said, with a kind smile, "is there anything you'd like to ask me?"

"Yes," I answered, rubbing my numb arm. "I've heard that if someone takes certain drugs they can beat your machine. Is there any truth to that?"

The polygrapher pursed his lips, then smiled, probably trying to decide if I was just a youthful go-getter or someone salting away potentially valuable information for the future. "No," he said emphatically.

"That's simply not true. If you take drugs before the test, the machine will register abnormal bodily responses rendering the test invalid."

I found myself irresistibly drawn to spy lore, which was infinitely more satisfying than the James Bond concoctions then available to the public. "And what about a report I read that said the Communists have a special school in Czechoslovakia for their agents on how to beat the polygraph?" I persisted.

In retrospect, it was incredibly presumptuous for a twenty-five-year-old TSD recruit to ask these questions. I hadn't even signed my secrecy agreement yet.

Still, the polygrapher was friendly and professional. "They might have," he allowed. "But we're fairly confident of our ability to detect any attempt at deception. It really is a matter of the examiner's skill, not just the technology of the machine."

Leaving the polygraph suite that morning, I had the feeling that this particular examiner would have been hard to deceive, and after a rigorous physical exam, I concluded that everybody I dealt with at the Agency had been thoughtful and professional.

Before heading home to Denver, I met with a CIA security officer, who briefed me on final procedures. "What have you told your family and friends about your interviews here?" he asked.

"I told them I was going to interview with the CIA," I answered.

He briefly considered this information. So far, there was no security problem: Thousands of people, including many of the brown-baggers on the commuter buses, openly worked for the CIA, assigned to logistics or administrative jobs, or to the Directorate of Intelligence, which analyzed the daily avalanche of information arriving from open and clandestine sources worldwide. The Technical Services Division of that time was part of the thinly disguised Directorate of Plans (later named the Directorate of Operations) and generally called the Clandestine Service. People working for this Directorate did not acknowledge their affiliation, but I was not yet sworn in. "It's fine for your immediate family to know you've had these interviews," the security officer said. "But you should tell anybody else that you didn't take the job offer from the CIA. Explain that you interviewed with other U.S. government departments while you were here and will probably accept a job offer with one of them if it comes through. We'll let you know which department."

THE POTOMAC RIVER ran seven miles from Chain Bridge near CIA Headquarters at Langley, through the limestone palisades separating Georgetown from Arlington, and under the bridge named for Teddy Roosevelt. Near the base of the bridge on the District side, the bluff rose again to encompass a small campus of unremarkable neoclassic limestone and brick buildings set in a quadrangle, the east side of which was the Navy's Bureau of Medicine compound on 23rd Street NW, fronting the State Department. In October 1965, the west side of this complex hung over a sheer drop down to an enormous, teeming construction site spanning a thousand yards along the Potomac, the future John F. Kennedy Center for the Performing Arts.

When I arrived in the quadrangle on a crisp fall morning, I stood for a moment quietly taking in the buildings. They had been the Headquarters of the OSS during World War II, taken over from the National Institutes of Health. To a fledgling spy, the OSS was legendary. The service combined both intelligence-gathering functions with covert action teams operating behind German and Japanese lines. The term "cloak and dagger" originated in nineteenth century Europe. It came into common usage in wartime propaganda films and press accounts describing OSS exploits.

Major General William "Wild Bill" Donovan, the OSS commander, was one such heroic figure. Awarded the Medal of Honor for World War I valor, Donovan built his wartime service into a global network, operating from bases in Britain, the Mediterranean, and South and Southeast Asia. He'd been a man of remarkable energy and innovation, attracting thousands of brave young Americans to his exotic unit. Beyond the Ivy Leaguers, there had been young infantry officers such as Bill Colby, who led a JEDBURGH team into Nazi-occupied France to rally resistance units after the Normandy invasion, and who later became Director of Central Intelligence (DCI) in 1973. Hollywood actor Sterling Hayden had run his own small squadron of armed fishing boats across the Adriatic to support Yugoslav irregulars pinning down several German divisions. Julia Child, who later taught America how to master the art of French cooking, served in the OSS in China. Popular lore also suggested that the OSS talent pool had included professional counterfeiters, con men, and safecrackers.

For me, the OSS represented initiative and bravery. Its operatives had engaged in daring acts of sabotage, subversion, and tactical deception against the Axis powers. Some had survived months behind Nazi lines, tracked by the Gestapo, living on their wits, audacity, and meticulous sets of personal effects and false documents contrived by the OSS predecessor-components of the TSD. Indeed, much of the structure, operational methods, and procedures of the CIA, created in September 1947, had evolved directly from the OSS.

It was a proud heritage. Years later, I discovered a photo of General Donovan in elaborate disguise, wearing a cleric's black cape and a primitive version of one of our more sophisticated devices, and I recognized that I had become part of an honorable legacy. Even on this fall morning, at the beginning of my new career, I felt a definite sense of professional kinship to my OSS predecessors. As I was entering on duty (EOD'ing, in Agency parlance), I sensed that I was taking up the baton passed to me by a team of older warriors in the long battle against totalitarianism.

World War II had been the greatest calamity in human history. The battered survivors on all sides had welcomed the end of the bloodbath. But perceptive leaders in the West, influenced by Winston Churchill, were forced to admit that a system of tyranny equaling Nazi Germany in its dehumanizing strength existed beyond the Iron Curtain. The KGB's predecessors, such as the notorious NKVD, did not herd people into gas chambers and kill them with Zyclon B; they packed them into cattle cars and sent them across Siberia by the millions to the frozen hell of the Kolyma mines, then worked them to death on starvation rations.

Although both Hitler and Stalin were dead by 1965, world Communism had established a terrifying record of conquest and enslavement. In the 1950s, when people in East Germany and Hungary had revolted against the puppet rulers whom the Soviets had installed following "liberation" by the Red Army, Russian troops and tanks massacred them. The Korean War had been a naked act of aggression launched at the behest of the Kremlin. Communist China, although moving away from its earlier alliance with the Soviet Union, was itself a vast, soul-crushing tyranny. East Germany was a police state that would have made SS Reichsführer Heinrich Himmler proud.

In fact, the Soviet overlords of "Democratic" Germany had been compelled to build the Berlin Wall in 1961—a glaring symbol of their failed system—to prevent a refugee exodus that would have depleted much of East Europe's most promising youth. That primitive, concrete-and-barbed-wire barrier, more than any other tangible entity, came to symbolize the geopolitical fault lines of the Cold War. As long as the Wall stood, the world was divided. This struggle was not merely an economic dispute; it was a war.

As the ten terrifying days of the Cuban missile crisis in October 1962 had shown, the Cold War could easily have flared into an exchange of thermonuclear weapons capable of incinerating cities across the world and poisoning the biosphere for millennia. There was also clear evidence that the Soviet Union and its surrogates, including the Cubans, were intent on spreading their form of liberation throughout the vulnerable wreckage of Europe's empires in the Third World. My generation of young adults, who came of age in the 1960s, were acutely aware of our tenuous world situation.

Was I proud to be enlisting, albeit in a seemingly minor capacity, on our side in the Cold War? You bet. Would I have preferred being a world-class painter living in New York or Paris, watching my kids grow up in a safe and peaceful world? Of course. But if I was going to become a cold warrior, I intended to be the best one I could.

The Technical Services Division front office had moved into the South Building from its site on the Mall since my initial interviews in April. Making my introductions among TSD offices that morning, I kept my eye out for any of the more exotic World War II OSS vets, but no one seemed to fulfill the swashbuckling image I had in mind. What I did find was a group of highly competent language specialists, artists, engineers, technicians, and versatile craftspeople capable of responding to bizarre requests from the field on a tight deadline. TSD also had its share of physics and chemistry Ph.D.s, as well as professional psychologists who could have thrived on any major American campus. The whole place seemed to resonate with competence and esprit de corps.

My orientation now complemented the brief introduction to this branch I'd received during my initial interviews. Many clandestine operations, I learned, drew heavily on authentication and validation capabilities, an elaborate way of describing the creation of alias identities and forged documents. TSD also "maintained" these identities through regular updates and renewals of documents, an unglamorous but vital facet of espionage.

But creating human identities was only part of TSD's responsibility. The Division also developed several types of seagoing vessels, large and small, a diverse array of aircraft, and a few basic spy cars, equipped with secret compartments for hiding officers or agents.

Some of TSD's most innovative mechanical inventions were probably the shipping containers equipped with automatic cameras dispatched from Hamburg to Hong Kong, via the Trans-Siberian Railway. When the thousands of clandestine photographs were developed, they revealed crucial details of the logistics system supporting Soviet nuclear weapons facilities in areas closed to foreigners.

TSD prepared other necessary spy gear, such as remote listening devices, the "bugs" of popular novels, and more sophisticated forms of "audio." TSD specialists traveled to aid clandestine officers on agent recruiting and rescue operations by crawling into attics to install audio devices, and meeting with agents to instruct them in the use of miniature spy cameras and invisible writing techniques. TSD later assumed responsibility for making compact radio transmitters that sent encrypted messages in almost untraceable bursts.

As former Director of Central Intelligence Richard Helms noted in his speech at the Agency's fiftieth anniversary ceremonies, at which I received my Trailblazer Award, there is no doubt that TSD work overseas can be hazardous. Helms cited three of the Division's officers, who were released in 1963 from Cuba's Isle of Pines Prison, where they had been held for two years and seven months. They had suffered repeated interrogation under torture, and survived only because the U.S. government managed to negotiate a prisoners-for-tractors exchange following the failure of the 1961 Bay of Pigs invasion.

One of the reasons they had lived through their ordeal was the quality of their alias documents and their ability to sustain their cover story during prolonged interrogation. These officers had been captured in Havana in 1960 while engaged in an audio penetration operation; soon after their arrest, it became clear that the new Castro government was a Communist-led Moscow proxy. Even in that notorious prison, however, the three managed to run a successful intelligence collection operation, working with the grim knowledge that they might be tortured to death (the fate of many of Castro's prisoners) had this brazen effort been discovered.

During its first fifty years, the Agency awarded only fifteen Intelligence Crosses, its highest decoration for valor. Three went to these brave TSD officers. Former DCI Richard Helms asked one of the surviving members of this team to lay the memorial wreath for all the CIA officers lost in action over the first five decades. Three of the fifty Trailblazers honored at this ceremony were former TSD officers. I am the only Trailblazer who holds the Intelligence Star for Valor, but several of my TSD colleagues have also earned this coveted medal.

When I came on board in 1965, TSD was organized under a Chief of Operations and a Chief of Development and Engineering. Since we were the technical arm of the Agency's Clandestine Service, we focused on assisting clandestine (that is, secret) information collection—material of interest provided by agents. CIA case officers assigned abroad work under official or nonofficial cover, as do the intelligence officers of many countries. A case officer might recruit an agent who operates alone against his government or goes on to become a principal agent, running a compartmentalized network of subagents. These agents might need technical support in the form of clandestine communications gear, spy cameras, and other specialized equipment.

Covert action propaganda printing, ranging in level of "plausible denial" by the U.S. government—from white to gray to black—was also a TSD responsibility. A white propaganda operation in the 1960s was merely promoting and packaging Western policy and culture, as with Voice of America programming. Gray propaganda might have involved writing and printing election campaign materials for a foreign political party friendly to the United States, or could have included planted news stories or editorial columns written for foreign press assets cooperating with the CIA.

Former veteran case officer Duane R. "Dewey" Clarridge has written about a black propaganda operation he conducted from his base in Madras, India. Clarridge used a non-Indian agent borrowed from another Asian country to convince the publisher of a local pro-Chinese Communist newspaper to run a series of increasingly virulent editorials, which eventually discredited the entire Indian Communist movement, both pro-Moscow and pro-Peking.

Black propaganda had been in use for decades by the time I joined the CIA. Soviet overseas intelligence officers, dating back to the KGB's predecessors, were experts at this nasty business. During the social and economic turmoil of the 1930s, the NKVD ruthlessly spread rumors of government corruption, backed up by forged documents, designed to inflame class divisions in Western Europe. Throughout the Cold War, the Soviets considered "disinformation" one of their important strategic weapons.

Equally cunning, American intelligence during the Cold War found it useful to encourage similar unrest among the workers at an arms plant in a Soviet-occupied country by circulating well-forged, ostensibly confidential official documents calling for an increase in labor quotas and a decrease in food rations. TSD/Graphics was especially adept at black propaganda operational support that made good use of our forgery capabilities.

Was this an "honorable" undertaking? In my opinion, it was justified. Our opposition practically owned the extreme left labor movement in a number of NATO countries and resorted to the same techniques. The KGB and its controlled intelligence services at the time were also actively engaged in "wet" operations—the assassination of political opponents in Eastern Europe and defectors who had sought asylum abroad.

I joined the ranks of TSD in an atmosphere of continual crisis. The watchwords were "Operations first," and "Can-do," meaning that the small TSD staff often had to respond with intense urgency to address breaking situations in the field. When an IMMEDIATE NIACT ("Night Action") cable requesting alias documentation for a fast-breaking operation hit DIRTECH (TSD) from an overseas station or base, people from Authentication and Graphics were often called in to work all night. In addition, we had to maintain over fifteen thousand alias identities at any one time.

That first day on duty, I was taken to the Art Department in a second-floor corner suite overlooking the courtyard and the Kennedy Center construction site below. There were about a dozen artists with various specialties crowded into a single bullpen. Franco, the Art Chief, and his deputy, "Rick," had small offices off the bullpen. We were located in only one section of the Graphics Branch, which occupied two of the Central Building's three floors.

Franco completed my introductions around the Branch. The offices on the first floor once housed the Institutes of Health's research animals and were known as the Zoo because some of the rooms still had cage doors. These rooms now held printing presses of all descriptions, a bindery, storage with climate controls for specialized paper, and alcoves for offset and letterpress platemaking. The process photo darkrooms, ink and textile labs, and the remainder of the World War II steel-plate engraving and die-sinking operations were spread haphazardly around the second floor.

During that war, printing near-perfect replicas of foreign documents—and enemy currency—had been a major part of the OSS graphics operation. Of the talented people who worked here, OSS legend Allen Dulles would later declare proudly, "Any intelligence service worth its salt should be able to make the other fellow's currency." That attitude was certainly a big stretch from Secretary Stimson's disgust at the idea of reading another gentleman's mail.

By the end of World War II, the OSS banknote engravers had completed the reichsmark and were almost finished with the Japanese imperial yen, which would have been used to debase the Japanese currency prior to the scheduled invasion in November 1945, had the atomic bomb not ended the war. But we could not counterfeit Soviet currency: Counterfeiting another country's money was officially an act of war, and the Cold War was not a declared conflict. In any event, by 1965, most of the world's currencies, including those of the Soviet bloc, were based on dollar reserves, so undercutting their value made no sense. Consequently, the two engravers remaining from those wartime days worked on specialized foreign identity documents, and occasionally designed and engraved the CIA's special citations and certificates for both officers and valorous foreign agents. In contrast to the artists, they had a huge workshop dominated by a massive, unused transfer press rivaling a steam locomotive in size, which had been used in making currency.

Because so many of the people in Graphics had been around since the war, they had seen thousands of interesting jobs flow through the shop. They had helped document hundreds of Allied agents and even anti-Nazi turnaround German POW agents, whom the OSS had parachuted behind enemy lines to locate buzz bomb and V-2 missile launch sites. During the Korean War, the Graphics shop had produced reams of bogus Red Chinese Army field orders in a covert action printing operation. Courageous agents working under American control managed to insert these documents into the Communist command system.

Each person in Graphics was expected to be a master artist in some specialty. In fact, many were bona fide eccentrics. Several had served overseas, and there was always a lot of talk on the floor about life abroad. Banter about traveling on the SS *United States* to Naples or attending Oktoberfest in Munich made the atmosphere in the bullpen very heady. The old hands would draw us new recruits aside and whisper, "The inequities of the job are more than made up for by the boat trips."

As I dived into my work, I became keenly interested in the operational purpose and goals of each project. I quickly discovered that this kind of curiosity was unusual among some of my colleagues, who were accustomed to producing a "job" skillfully on a very tight deadline and had little interest in precisely how the documents they were so artfully replicating would ultimately be used. Such details might have been too disquieting to hear.

Certain Graphics people did not always mesh with their TSD colleagues. Although I had been hired as a GS-9, others in the Art Department worked on "wage board" standards and had their pay raised when the unions at the Government Printing Office got theirs. This gave some of the employees a slight blue-collar attitude. In fact, my being hired on the professional GS scale was an attempt by TSD managers to undercut the union mentality among the photographers and pressmen of the Central Building. Certainly Ian Fleming never wrote about such pedestrian concerns, and neither did the more sophisticated John Le Carré. However, this was the gritty reality of American intelligence.

As it turned out, there was no way management was going to convert the pressroom workers to the Civil Service scale. Working under wage-board guidelines, some members of the Graphics Branch already made more than the Chief of TSD, who was a supergrade GS-18. This was because the wage-board system paid generous overtime, including double-time for weekend jobs. Civil Service employees had much more stringent overtime provisions.

After the standard orientation courses for new employees, I volunteered for area studies classes and strategic seminars in order to widen my perspective. In these courses, I became acquainted with people throughout the Agency, including the CI (Counterintelligence) staff, headed by James Jesus Angleton, the legendary "Gray Ghost." Angleton had been badly burned by the treachery of the Soviet's British agent Kim Philby and was obsessed with the idea that the CIA had been penetrated by a highly placed Soviet "mole." Unfortunately, his theories were so convoluted, and his methods so Byzantine, that he was forced to resign in 1974. Besides the occasional oddball, however, I found members of the CI staff relatively free from paranoia.

The nuts and bolts of CI training was fascinating. We were shown

photographs and diagrams of captured Soviet bloc listening devices em-
bedded in a diplomat's shoe, in the spine of a book presented to an
American officer overseas, and even in the Great Seal of the United
States, which the Soviets had offered as a "good will" gesture to our
ambassador in Moscow. I came away from these classes with the knowledge that the shadow battles on the Cold War's espionage front would
be fought relentlessly, with great cunning, for years to come.

With this in mind, I tried to absorb as much as possible about the
actual impact of the Graphics operation. I learned that our products
had to withstand the severe scrutiny of hostile services that had been at
this business longer than we had. Much of what we produced was agent
and officer documentation, which went beyond simple identity cards to
include entire families of documents—birth certificates, school records,
military service dossiers, the full gamut of paper needed to authenticate
an alias identity, or "legend." Early in my Graphics assignment, I discovered that the paramount challenge facing us with respect to documentation was embodied in the question: "How good is good enough?"
The best way to answer that question was with another: "Would I use it
myself in hostile territory?"

By the 1960s, the Agency had learned the painful lesson that running
agents behind the Iron Curtain was a lot more difficult than working
with resistance groups in the Nazis' Fortress Europe in World War II.
Unlike the OSS, we could not parachute an agent into Hungary, Poland,
or Rumania to be supported by a sympathetic and effective underground. That sad lesson was again learned with the Bay of Pigs disaster.
An armed and organized resistance simply didn't exist in Soviet bloc
democratic republics as it had in countries under Nazi occupation.

On many early cross-border operations into the Soviet bloc, agents
were "rolled up"—seized and perhaps never heard from again—or "dou-

bled back"—taken under Soviet control—or simply disappeared, never having intended to spy for the West in the first place. The extreme security systems of the Communist governments (the East German Stasi being the most notorious example), coupled with decades of indoctrination, made entire populations suspicious and willing to spy on neighbors or coworkers for the collective good. Although the fascist governments had never been able to penetrate the fabric of the societies they occupied, the Soviets, working with security services such as the Stasi, virtually controlled the societies of their satellite empire for decades.

This made running intelligence operations in those countries extremely difficult and hazardous. We now know, for example, that the Stasi operated vast networks of informants, many of them reporting on one another, often family members, even husbands and wives. Under Soviet guidance, the Stasi trained "fraternal" intelligence organizations, such as Cuba's DGI. During my career, the Stasi and the DGI were responsible for some of the most successful double-agent operations run against Western intelligence. We learned a lot more about those relationships after the fall of Communism.

The first decades of the Cold War taught the Agency in general, and TSD in particular, some harsh lessons about cross-border operations. We had been able to reproduce sophisticated and complex materials such as international travel documents and internal identity papers. Only rarely, however, could we "backstop" them by having trustworthy agents in place in the issuing offices to plant the file supporting the false items. This meant that if one of our agents was apprehended because of his flawed demeanor—or Murphy's Law—his or her false documents could be exposed within hours. Therefore, we had to find out much more about the minute practical details that legal travelers to those countries faced in order to move freely without undue suspicion. For example, did a Latvian soldier in the Soviet Army need special military and civilian papers to ride a train to visit his "aunt" in the Russian Republic?

Our tradecraft was slowly evolving to become more adept at building neutral, "third-country" cover legends and aliases that were more innocuous and difficult to detect, even without backstopping. This level of spycraft requires years of patient preparation and had long been used by sophisticated services such as the KGB in their "illegals" (spies with nonofficial cover) program, which infiltrated hundreds of bogus "refugees" from Europe and South Africa after World War II. But the Agency was still mastering the "neutral" alias approach when I came on board.

I absorbed much of this lore from both training courses and colleagues. What I found especially fascinating was the combination of intricate sophistication and massive deployment of manpower that the KGB expended on the espionage front. Beginning with the Cheka in the early days of the Revolution, and which was well established in Britain by the 1920s, the Soviets steadily deployed spy networks throughout the West. Their recruits had penetrated the British intelligence services before World War II, had several key agents recruited in the OSS hierarchy and, as we now know, had crucial spies in the Manhattan Project, which designed and built the atom bomb. Their networks were run by known KGB officers operating out of their *residenturas* under diplomatic cover, and by an unknown number of illegals, such as Colonel Rudolf Ivanovich Abel, who operated under deep cover without diplomatic immunity.

In 1946, Abel had made his way to Canada, documented as a German refugee, then simply slipped across the border into the United States. Living in New York under the alias Emil R. Goldfus, he controlled a large KGB agent network that stretched from Canada to Panama. He was discovered only when a drunken subordinate defected to the West and he was traced to an artist's studio rented in his "real" (i.e., alias) name—a major breech of tradecraft. Today, knowing so much more about the KGB's practice of redundancy, we can only speculate how many other Colonel Abels there were who didn't make such blatant errors.

As I settled down to my work in the Technical Services Division, I experienced a growing sense that, like the agents and their case officers working overseas, I had left the world of normalcy and entered a domain of shadows.

3 ■ Onto the Shadowy Battlefield

[As our] spies we must recruit men who are intelligent
but appear stupid; who seem to be dull but are strong
in heart; men who are agile, vigorous, hardy and brave;
well versed in lowly matters and able to endure hunger,
cold, filth, and humiliation.

  —Sun Tzu

Technical Services Division, 1965–66 ■ My first practical assignments
in TSD Graphics Branch required producing the elements of personal
alias documents. These projects went far beyond simply replicating official papers. I learned to work with linguists and area specialists who
had studied foreign travel and security controls, often through debriefings of third-country bridge agents (CIA agents building a "bridge" into
the target country), and were now much better equipped to design the
"families of documents" our case officers needed to operate around the
world.

It was tense work. We never knew for sure if we had identified all
the security traps a foreign government had hidden in its travel papers

or identity cards. This uncertainty caused anxiety among some Graphics people, who worried that agents using our documents might be rolled up. For some, the anxiety never eased, especially when agents entered East Germany and Czechoslovakia. Our counterparts in their security services were always inventing more elusive features and traps, so their documents were becoming increasingly more sophisticated.

Since two of the traditional agent entry routes into Eastern Europe passed through East Germany and Czechoslovakia, it was no coincidence that their immigration control procedures and inspectors were among the most highly skilled and disciplined in the Soviet bloc. Usually, however, Western intelligence could work around this problem by entering or exiting Eastern Europe through softer avenues of approach, such as Hungary or Poland. Once behind the Iron Curtain, a handler might meet an agent traveling east for a rendezvous, rather than west, and thus arouse less suspicion.

But to support such operations, TSD had to constantly collect, analyze, and probe immigration and internal security documents, as well as the procedures governing their use, to develop effective "recipes" to penetrate these controls. Personal identity documents demanded special research and attention to detail, as I discovered on my earliest tasks as a TSD graphic artist. For example, a simple reproduction of an East German border-crossing cachet (typically, a rubber stamp) had to be regarded as a sophisticated security instrument. I remember well when Franco presented me with a stack of travel documents all bearing the immigration cachets of East German railroad immigration police who had inspected travelers from Budapest to Berlin.

"Study them carefully," he said. "Then tell me what you've found."

Under high magnification, I discovered that the Roman numeral V and the Arabic numerals 7 and 3 each had miniscule flaws, as if the ink had not adhered to the passport. In one or two cachets made on the same day, this was understandable. But these samples from case officers' and agents' travel documents accumulated over two years revealed subtle but extremely effective security controls in these impressions. If we tried to replicate this cachet without the intentional flaw, anyone using it in an alias document was likely to be rolled up.

That was just the rudimentary end of the security document business. TSD artists took extreme care in identifying the type fonts used on genuine documents to determine exactly which foundry had produced a particular typeface. The subjective quality of a rubber stamp or printed impression was equally important. Rubber stamps were often pounded, especially by Communist bureaucrats, and the stamps became worn. The riskiest thing we could have done to an agent would be to issue him documents bearing stamp impressions that were too crisp. When we actually applied these cachets, we went to great lengths to get the right look, repeatedly inking the stamp in a rhythmic motion to imitate the pressure that a busy immigration inspector would have used. Composing a page in a travel document with several forged "chops" stamped at slightly different angles—some fainter, some bolder—resembled the act of painting.

Whenever I worked on an immigration cachet, senior artists constantly checked my progress, patiently advising on the right pressure to apply with my fine-tipped crow-quill pen as I stippled ink onto the photo master page, which would be rephotographed and reduced to size. Rick or Franco always carefully inspected my work before sending it down for photo/plate processing. Only after I mastered the cachet jobs was I allowed to handle a "repro" of a secondary document, such as a national ID card or a military driver's permit. These assignments were more sophisticated and required closer scrutiny and deeper knowledge of the printing process involved so that I could "pick the fly shit out of the pepper." Fulfilling these demands was not fun. The exacting work entailed neck-wrenching hours bent over the drawing board, gazing through a powerful industrial magnifying lens and wielding a variety of sharp tools with surgical precision.

In spite of the agonizing strain, I have never regretted that apprenticeship in the Graphics bullpen. It taught me that real-world intelligence is not a game, but a highly demanding trade with a terrible price in the event of failure—a human life.

When the shop was assigned a major document reproduction job, the whole team rallied to the effort, with the most senior artists drawing the toughest tasks. But even junior artists like me had to shoulder unusual responsibility and exercise astute judgment, because mistakes made early in the process were often expensive and time-consuming to repair if discovered later.

Harrowing in a different way were the rushed passport issuance or validation jobs. For example, a case officer in the Far East might receive word of a diplomat of intelligence value wishing to defect, which would require a complete set of alias documents for the man and his family, to be hand delivered within a few hours. If I, the artist reproducing and applying the forged document entries under this deadline, made a serious error, it could cost the life of the defector. Later, I applied lessons learned in my years at the Central Building during overseas operations in which the subjects were agents or defectors under hostile pursuit—people who would have been imprisoned or executed if captured.

I have to admit that I was absolutely engrossed by this work. It was not just the technical challenges, but the full scope of clandestine operations that intoxicated me. I was confident that I could master quickly the skills required in the overseas program, but realized I was woefully low in the pecking order for a coveted assignment abroad. I decided to demonstrate my capacity to work hard and tackle mundane jobs while seizing every opportunity I could to contribute to the most demanding projects.

Soon after I came on board, I began carpooling from the Virginia suburb of Falls Church with "Wilhelm," one of the most talented artists in the bullpen, and managed to absorb a great deal of the subtle intricacies of our craft directly from this veteran. Wilhelm, his wife, Nancy, and their children lived near Karen and me, so this older couple became our unofficial mentors. This cooperative atmosphere was typical of the Agency, which remains a very collegial business. I have heard that other government departments are notorious for backstabbing and careerism among officers scrambling over each other's backs like Maryland blue crabs in a basket. My perception of the culture of the Agency, however, was similar to the camaraderie encountered in elite military units.

I quickly developed the habit of eating lunch at my drawing board, which, I noted, was also Wilhelm's custom. Sometimes during lunch hour, a hot job requiring immediate attention would be brought in. Perhaps a courier was leaving that afternoon, and a TSD authentication officer in the field needed one more piece of agent documentation to complete the legend or even assist in an exfiltration under hostile conditions. On these jobs, I worked with Wilhelm or one of the other experienced artists, which was both exciting and very instructive.

Roy and Doris, artists who had been in Graphics the longest, told me colorful stories of their early years in TSD, when the Agency had been smaller and the workload less pressing. In those good old days, some of the European-educated artists would bring in their violins, clarinets, and oboes and entertain the second floor with chamber music. Other artists had set up their easels and worked on abstract paintings or landscapes, rather than on the mundane but exacting tasks of espionage graphics.

I can attest that all had changed by the time I arrived. We had to account for every hour of our working day with a personal job chart. Rick, the deputy chief, picked up "the hours" each afternoon and entered them in a ledger. While others chafed under this sweatshop approach to production, I found it challenging to see how many jobs I could complete as quickly as possible. In a way, I felt as if I was in Denver, hustling for myself to make ends meet.

One of Rick's habits rubbed me the wrong way. He always arrived earlier than everyone else and inspected their work in progress, often marking perceived errors in the art work with small blue arrows, thus forcing the artist to begin all over again. Rick, I soon learned, had a mean-little-kid streak, and reminded me of the journeymen plumbers for whom I had once worked—grown men with little sympathy for the new kid on the block. They would send me to the truck on a wild scavenger hunt to find the "left-handed pipe wrench" or the "glass stretcher."

Then there were the pressmen and photoengravers who clung stubbornly to their crafts' union pride and were eager to put an apprentice artist like me to the test. One afternoon, I entered the pressroom and found Ralph mixing a large batch of ink on a slab of polished limestone, an art that dates back to Medieval Europe.

"Ink gets hot when you mix it like this," he muttered. "C'mere and feel it."

When I took the bait and held my hand over the ink, Ralph grabbed my forearm and pressed my palm into the gooey mess. "Feels hot, don't it?" he said with a laugh, tossing me a rag.

Tricks like these were all part of the initiation process. If I could put up with this type of prank, I knew I could handle more stressful situations.

UNLIKE OTHER GOVERNMENT departments, there were no guarantees of advancement in the CIA. Promotion to a higher grade was not based on time in service but on competitive evaluation. We all worked at the "pleasure" of the Director of Central Intelligence and could have been fired without protracted Civil Service appeals. But I also realized I could advance at my own pace if I showed initiative and rapidly improving skills, keeping an unwavering eye focused on an overseas assignment as an artist.

But my time in the Graphics bullpen was not all drudgery. On payday, the whole crew went out for lunch to favorite haunts such as Peugeot's on Connecticut Avenue or One-Step-Down on M Street. These restaurants were frequented by most of the Clandestine Service, not just people from TSD. You could spot the CIA people immediately; they kept their neck-chain ID cards hidden inside shirt or blouse pockets. It was possible that the KGB bloodhounds from the Soviet embassy on 16th Street also bellied up to these bars to eavesdrop two Fridays a month. But I never once heard an Agency employee even whisper a fragment of operational information in public.

I soon learned that the Agency was an army that traveled with more than one martini in its stomach. Back at the office, it was an Agency tradition to haul out a bottle of scotch from a desk drawer late on Friday afternoons and discuss the progress of the week's work in a ceremony called "the vespers," which dated back to OSS days. But again, I was impressed by the ability of senior artists and officers, especially those who had served overseas, to hold their liquor. In that regard, they were not unlike career military officers I would soon meet in the Far East.

The very nature of working in the Clandestine Service tended to bring not only employees but also their families together. We were not allowed to talk about our jobs to people outside the CIA, so Karen found it especially comforting to have close friends such as Wilhelm and Nancy, who had been in the Agency for years and had served overseas with their kids. Agency employees often socialized with each other, in part because many of us were working under "nominal" cover at other government departments.

My cover legend was tested during my first year in the CIA at a party thrown by a neighbor to celebrate his promotion to air force lieutenant colonel.

"So where do you work?" another guy from the block asked me.

"For the government," I replied.

Unfortunately, that answer was always a dead give-away to the amateur spook hunters thriving in Washington. "What department?" he pressed.

I responded with my nominal government agency, but he claimed to have friends there and asked me for the exact location of my office. Frustrated, I provided the floor and room of my cover job but a sly, grin spread across his face.

"That's the mail room," he proclaimed. "I service the Xerox copy machines in that building." He lowered his voice. "You work for the *Company*." All that was missing from the encounter was a conspiratorial wink. After the James Bond craze, it seemed that everybody wanted to be a spy.

I managed a weak shrug and turned away. I did learn later to avoid such confrontations altogether. There is no perfect answer, but a person can become an artful user of cover—and humility is a powerful tool in the spy business.

Although the FBI ran a pretty effective counterintelligence operation, its agents could not be everywhere. It was simply safer not to advertise that someone like me was the new boy in TSD's Graphics Branch. Our cover legends mostly served to divert the paper trail in the public record away from the Agency. Therefore, "living" a legend was not only the discipline and duty of the officer, but of his entire family as well. Once a CIA family's cover was established, in many cases with alias names, the Central Cover Staff at Headquarters became actively involved in their personal lives. The Agency handled credit and real estate references, as well as banking queries. It even vouched for parents' employment on their children's college applications. Karen and I were learning to use cover effectively in the States, preparing us for overseas assignments.

After more than a year on the job, however, I started to worry that I might not make it overseas any time soon. The line for artists' jobs at TSD bases abroad seemed to be getting longer. Finally on an afternoon in May 1966, I heard that TSD was putting together another TOPS class.

That night after dinner, I took Karen aside. "I'm going to toss my hat into the ring for the class," I vowed.

A week later, I was in Phil's office. "You weren't selected for the TOPS class," he said, with an edge to his voice. "You know the basic requirement for the program is a previous overseas assignment in your primary skill."

"I'll keep trying any way I can," I said. "But if I can get overseas as an artist, that's fine with me."

"Artists are a dime a dozen!" he said. "We can always get another artist."

This statement was foolish, and I challenged him. "It takes a long time to train a TSD artist, and I'm proud to be one. I'd just like to work overseas. That's one of the reasons I joined the Agency."

Phil was clearly annoyed by my persistence. "Well, Rick is going to the Far East this year, and next year it's Joe's turn. You'll just have to hang in there."

Rick was scheduled for his assignment as a senior artist in June, and no one was due back from Europe for a while. I was faced with a stagnation for at least two more years in the bullpen. Leaving Phil's office, I was more determined than ever to persevere and press for as much training as I could get.

I soon found an opportunity to speak with Dr. Gottlieb, now the chief of TSD. The TOPS program was originally one of his concepts.

"How are things going over in Graphics?" he asked.

"Fine," I said. "But I did apply for TOPS training this year."

"I know," he said kindly, "and I hope you keep trying."

Instead of departing like a schoolboy, I held my ground. "I'd like to get the RCA electronics course out of the way." This was a correspondence course required of all TOPS graduates. If I could complete it in advance, I'd be ahead of the curve.

Phil was not amused when he learned that Dr. Gottlieb had ordered the course for me. But I told Franco that I needed to advance in the Agency. Trying to raise three young children on my salary wasn't easy. Karen and I had furnished our small rented house after scouring swap shops and garage sales. Our evening "cocktail hour" meant splitting a single can of beer, and once a month on payday, we took the kids to McDonald's. Although I had put a workshop and an art studio in the basement, we lived from check to check. The prize money I'd won from two best-of-shows at local art exhibitions had come in handy, but had quickly disappeared.

Then, during the week before Christmas of 1966, Franco called me into his office. "You're the next artist in line for a PCS overseas." PCS meant permanent change of station. He smiled broadly. "But you'll have to hustle through intensive training to join Rick in the Far East next June."

"How did this happen?" I was flabbergasted.

"You've been doing a great job, and management thinks you're the best qualified. Besides, Rick requested you."

Overseas! I had trouble containing my excitement and couldn't wait to tell Karen. But as I entered the bullpen, I saw Joe watching me. I'd managed to surpass him not through backroom deals, but through hard work.

He smiled and offered his hand. "Congratulations. You deserve the assignment."

"Camp Swampy," Virginia, March 1967 ■ "The Farm" was a sprawling ten-thousand-acre facility of prime tidewater real estate near Williamsburg, Virginia, where I took the OPS FAM (Operations Familiarization) Course. It was the place where Agency officers headed overseas learned their tradecraft. Although shorter than the Operations Course that case officers had to pass, OPS FAM also had the intensity of elite military training. For six weeks, we would be in isolation, learning to walk in the shoes of agents and their case officers during CIA field operations. To heighten the somewhat spartan atmosphere and the serious purpose of the course, we worked in alias and wore military uniforms.

"Arthur Masters," the man in Army fatigues behind the counter said, staring at me. It took me a second to remember that this was the alias I'd been assigned. Thirty-some students were standing in line at the uniform warehouse, where the bus had dropped us off on this first day. The supply clerk handed over a pile of highly starched khaki he'd assembled based on the size slip I had submitted.

"What size boot?" he barked.

"Nine and a half," I replied, an octave too high. I could have been a seventeen-year-old private in Marine boot camp.

The bus took us back to the BOQ area, where we changed into our khaki uniforms and assumed an even murkier identity than our aliases. Since there was a case officer OPS course already under way, the sense of being part of a much larger clandestine effort was enhanced.

Our lectures and tradecraft demonstrations took place in Arena A and Arena B, modern brick buildings hidden by trees in this "school for spies" that stretched across several colonial farms. There were also older buildings, ranging in style from a Virginia planter's manor house to 1930s tar paper. The instructors and staff lived in these old farmhouses, and the nicely restored older buildings were available to senior CIA officials for weekend getaways and private conferences.

The Farm had also inherited several graveyards from the previous residents. Although outsiders were not allowed into this severely restricted training base, there was special consideration made for grave visits, during which students were kept out of sight. With hunting forbidden, the Farm was overrun with deer, including some spectacular bucks, and coveys of quail shielded from the local gentry with their Purdy shotguns.

Although it has not been generally discussed, the Clandestine Service was not an Equal Opportunity Employer, even after Affirmative Action. Several of my fellow students at the Farm were women, and only a few were ethnic minorities. The CIA was still dominated by an "old boy" ethos that impeded flexibility. Later, as a manager and senior officer, I would help reverse this situation.

We weren't marched around by drill instructors inflicting abuse, but we were expected to walk fast and toughen up in general. Unlike diplomacy, espionage is a mentally and physically demanding business. Overseas, we learned, when the diplomats were falling into bed relieved after a long evening of cocktails and dinner, the case officers, who had attended the same parties, were out on the street, heading to their meetings or sending late night cables. CIA employees could be easily identified from the cars parked at the office on weekends.

I found the tradecraft instruction even more fascinating than my training in Graphics. Somehow, over the course of only six weeks at the Farm, the instructors managed to provide detailed practical classes on writing cables and reports using manual typewriters; conducting and evading surveillance; making and placing "dead drops" (caches where material may be left for others); casing a site; debriefing a source quickly and efficiently; preparing and issuing a secret agent's communications plan; and all the other taxing requirements involved in spotting, recruiting, and handling an agent or "asset."

Certain classes at the Farm remain vivid in my memory even after more than thirty years. Clandestine communications is one. We were taught to use the one-time pad, an ingenious enciphering method dating back at least to World War I. As the name implies, one-time pads were meant for a single use. The ciphers were composed of thousands of blocks of digits, selected by complex mathematical formulae to ensure maximum randomness. In the 1960s, our best computers were used for this task; today, one-time pads are produced by the most advanced computers in the world.

One-time pads were printed in only two copies, one for the agent or man in the field, the other for either the case officer who issued the pad or for use at Headquarters. To encipher a message, the agent or case officer selected a page on the pad and converted the clear text into the cipher digits, which could then be transmitted in several ways, even by commercial telex or telegram in countries where that practice was permitted, or more often through clandestine means.

The person deciphering the message simply went to the precise page and line of the pad to transfer digits to letters. For even greater security, the commo network could double encipher the message from one page of the pad to another.

But the crucial security factor in this system was that the pages used were *always* destroyed after the enciphering and deciphering at both ends. Pads came in many sizes; some were tiny Torah-like scrolls on specially treated "flash" paper that would flare up and instantly burn at the touch of a match. Other one-time pads could be concealed within the covers of books or notepads, or as latent images on a variety of substrates or carriers. In the Agency, TSD was responsible for continually upgrading both the efficiency and the security of one-time pads.

Case officers could send their agents these enciphered messages by radio without resorting to the old open-broadcast voice codes of World War II. Standard short-wave transistor receivers, available worldwide, could be modified to receive these signals on the edges of normal broadcast bands. Naturally, the opposing security forces had equipment to intercept these messages as well, but what they recorded was gibberish.

To reply using a one-time pad, an agent could encipher his message and leave it in a one-time dead drop "mail box" serviced by the CIA on a precise schedule. Or the agent could employ secret writing techniques (some involving invisible inks) in messages mailed directly or delivered through an accommodation address or "cut-out" (an intermediary step separating sender and recipient) such as a lateral series of agents, or "live drops."

Learning the art of street work in a hostile environment was equally intriguing. Mr. O'Brien, our instructor at the Farm, was an experienced case officer who had operated in many of the spy capitals of the world, including major Soviet bloc cities. He was the quintessential "little gray man"—short, balding, dressed in a drab business suit with a cheap tie and nondescript shoes. His raincoat and blue porkpie hat were equally unremarkable, but the cap was the only hint of color any of us ever noticed about him. He was the kind of person you might never notice in the first place, someone whom you would forget two minutes after he passed you on the street or rode with you in an elevator.

"Always assume the hostile service is omnipresent and alert," he cautioned us. Indeed, O'Brien presented evidence that the best Soviet bloc security services and their protégés in the "nonaligned" third world devoted substantial resources to street surveillance. They often ran internal exercises, keeping their people at a high state of readiness, especially the Egyptians, Algerians, and Indians.

He taught our OPS FAM class surveillance techniques and showed us several films that illustrated the basics. Then we were allowed to reclaim our civilian clothes and were turned loose on the sunny spring streets of Richmond in a game of hide-and-seek, trying to maintain discreet surveillance on Mr. O'Brien, while he tried to both detect and evade us.

One day I was assigned to a three-student, ABC-formation surveillance exercise that attempted to follow O'Brien from the slope of the State Capitol building through the downtown streets, just as the lunch-hour crowd was spilling onto the sidewalks. We were going to play a real game of fox-and-hounds: He knew he was under student surveillance, but he hadn't been briefed on the composition of our team.

"Little gray man" that he was, he blended in seamlessly with the crowd, wearing a beige raincoat on this warm March day of sunshine and showers. With myself as the tail man and my two teammates running point and wing, we formed an arrowhead of "pedestrians" surrounding the target, which rotated in a precise order. O'Brien's beige raincoat and the rumpled blue cap were easy visual cues, which I tried to keep in my line of vision, along with the dark blue sportscoat of my point man, who was weaving through the crowd ahead of me, trying to maintain a safe distance from O'Brien.

He appeared to offer no real challenge to our pursuit; although moving at a comfortable pace, he was certainly not trying to outdistance us in any obvious way. In fact, he paused to buy a newspaper on one corner, then cut back across the street to duck into a Walgreen drug store and quickly buy a pack of Winstons. The point man had easily detected all these movements.

I glanced across the street at our wing person as she strolled along, a nonchalant window-shopper. So far, so good. We had followed O'Brien more than four blocks and had been told to expect his "brush pass," a brief encounter with an agent, a half-mile from the Capitol. Only two blocks to go, I thought, feeling triumphant.

But when I looked ahead again, I no longer saw O'Brien's blue hat bobbing in the crowd. I caught up to the point man, who now stood transfixed on the sidewalk.

"Where did he go?" the point man whispered in anguish.

"I thought he was right there ahead," I answered, looking at our wing across the street, but she just shrugged, signaling she didn't have a clue either. We paused in the flickering movement of the sun and shadows, fledgling bloodhounds who had lost the scent. Then I scanned the street ahead and saw O'Brien emerging from a department store. He paused briefly to put on his cap and raincoat, and strode purposefully down the street once again. Slippery rascal! We'd failed the exercise because the pass must have occurred in the crowded department store, when O'Brien and the agent had "brushed" by one another and exchanged packages, or dropped a message in each other's pocket or shopping bag, undetected by our surveillance team.

In the critique of that exercise, O'Brien recited the exact composition of our point, wing, and tail formation, describing in detail what we wore and our blatant actions, which we had thought had been so subtle. While we had been trailing him, he had conducted his own skillful SDR (surveillance-detection run). Pausing to buy the newspaper, he'd spotted both the point man and myself freeze on the sidewalk like startled deer, then try to look nonchalant in vain, as we treaded water to maintain a safe visual distance. As he crossed the street to buy cigarettes, O'Brien had caught sight of the wing, who had automatically turned her back on the target, and watched his reflection in the window (an old gangster movie ploy). None of us had "gone with the flow" and kept up with the crowd.

With humor and patience, O'Brien explained how we should have handled the exercise. "You shouldn't have stopped moving when I bought the paper," he told the point man and myself. People freezing abruptly in a continuous stream of pedestrian traffic are easy to spot. "Don't turn your back suddenly while you're window-shopping," he told the wing, suggesting that she should have continued walking right past the Walgreen, then positioned herself down the block. When she protested that he might have slipped out the back of the store, he smiled. "How many Walgreens do you know that let customers out the back door? That's why exact area knowledge is vital."

Once he was certain the surveillance team consisted of only three, he was able to shelter himself in the crowd, quickly fold his cap into his raincoat, and, in a matter of seconds, slip naturally into the department store, which was crowded and had multiple exits. But we had been so fixated on his blue cap and raincoat that we had failed to grasp a fundamental principle of surveillance: A good street professional could change color like a chameleon. "Follow the target person," he reminded us. "Don't focus on their superficial appearance." This was a lesson I never forgot.

■

THE ROOM IN the BOQ at the Farm was lit only by a small desk lamp, and both the blinds and curtains were drawn tight. I sat waiting on the military-issue desk chair for what seemed like hours. Finally, there came a faint tapping at the door, and then a disheveled little man darted into the room.

"Can you close the door, please," he asked in a heavy Slavic accent, casting a furtive look down the corridor. Very quietly, I latched the door and invited the man to take a seat.

"What can I do for you?" I asked.

His hands were shaking as he lit a cigarette. His eyes were blood-shot, and I could smell alcohol. In just a few minutes, the pressing reality of the situation had struck me full force. "I'm the Second Secretary of the Embassy of the Republic of East Sorbornia. I am here to request political asylum."

He leaned forward, eagerly searching my face.

The man possessed such impressive acting skills that I almost forgot my instructions for handling this type of situation. A "walk-in" defector of any value was a prize in the spy business, but sometimes these plums came on thorny branches. A provocateur could launch a propaganda scandal by claiming to have been kidnapped by the Americans, or a potentially valuable defector might become intimidated after a bungled first contact with the Agency.

"What have you got to demonstrate your bona fides?" I asked, perhaps too harshly.

He reached into his suitcoat. "I have my diplomatic identity card, and I have brought along a confidential dispatch and a cable from my embassy that might interest you." He handed these documents over to me.

The identity document looked genuine. The diplomatic correspondence (in English for this exercise) was stamped "Most Secret," and the text appeared very intriguing, based on my quick scan.

"Do you have any information about an impending attack on a person or place of interest to the United States government?" This was the second item on my mental checklist—the mandatory Early Warning question. After all, he might have defected because war was imminent.

He frowned. "No. I have no information about that." If this man was lying, I thought, his performance was worthy of an Oscar. I was convinced he needed help.

"Who else knows you're here?" I asked.

"Only my wife and son, Sasha. They are waiting for me down the street, in the café on the corner."

"Can you go back to your embassy and remain there a while longer?"

Our lecturers had emphasized the importance of this approach. Some of the best intelligence sources in history had been volunteers who were actually able to remain "defectors in place" for years following their initial walk-in contact.

"Yes," the man mumbled. "I think I can stay there safely for several more weeks."

Bingo! My part in this simulation exercise had been successful, and I was relieved.

One of my most important tasks as a TSD officer overseas would be to maximize defector readiness for safe exfiltration out of our operational area. If U.S. Intelligence could keep a man in place long enough to collect valuable information, then successfully exfiltrate him and his family (perhaps after planting some audio souvenirs in his embassy), we would have earned our pay.

Exfiltrations were the most exotic application of my specialty as an espionage artist. One of the most effective ways to persuade a defector-agent in place to walk this dangerous tightrope was to assure him that there were convincing alias identities and exfiltration disguises for immediate use if necessary.

Another TSD support tool for walk-ins or an agent in place was prepackaged standard espionage tradecraft—language cards used to describe clandestine communication plans; training in dead drops; and procedures to make debriefing sessions possible at safe houses.

Perhaps the most important lesson I took away from my six weeks at the Farm was the visceral knowledge of what it was like to walk in the shoes of agents and their case officers.

The final lecture in the OPS FAM course was delivered by George Kesvalter, the Agency's only supergrade case officer, who had earned his rank handling Soviet GRU (military intelligence) defectors-in-place, including Colonel Oleg Penkovsky. Penkovsky had provided some of the most valuable intelligence of the early Cold War before he fell prey to faulty tradecraft and was subsequently arrested and executed in Moscow.

Despite advances in technical intelligence, including the growing use of high altitude reconnaissance (PHOTINT) and communications intelligence (COMINT and ELINT) satellites, Kesvalter stressed that human-based collection (HUMINT) espionage remained as much an art as a science. Common sense, and the ability to analyze character quickly and decisively, were the intelligence officer's greatest assets. Kesvalter emphasized his points through a series of captivating, sensitive accounts of successful penetration operations, in which case officers perceptively assessed their agents' potential strengths and weaknesses and ultimately guided them to safety.

After the lecture, we followed Kesvalter to the bar in the Student Recreation Building, where he continued to regale us with anecdotes and advice. We hung on every word as he reminisced aloud, reciting chapter and verse on the pedigree and bloodlines of the "opposition," the KGB. The notorious agency of the enemy had evolved from the MVD and its predecessors all the way back to Lenin's first spymaster, Feliks Dzerzhinsky ("Iron Feliks"), one of the bloodiest murderers of the twentieth century.

Two days later, my OPS FAM course had its last Friday night party at this occasionally rowdy bar. Thousands of Agency officers had passed through the Farm and sweated out their training here; many had soothed their frazzled nerves with alcohol. That night, the mood was jovial and celebratory: Most of us had done well. After several rounds of toasts, we launched into a game of "Carrier Landing," in which each competitor flung himself down the length of four tables drenched in beer. It was a messy and noisy game, but a great deal of fun.

Unfortunately, the stress of overseas Agency work sometimes led to alcohol problems. In those days, some of the old hands, who had spent years on Soviet bloc streets or in the steamy capitals of Southeast Asia, finished their careers as staff on the Farm, far removed from Headquarters. But that practice ended years ago. Today, the Farm is staffed by many of the Agency's brightest and most innovative officers, who realize that educating promising overseas officers is one of their most important assignments ever.

South Building Laboratory, Midnight, May 24, 1967 ■ I held my breath as I peered through the stereomicroscope at the ragged strands of paper fibers. Hunched over the lab bench, a scalpel poised in my hand beneath a high-intensity light, I felt light beads of sweat trickle down my back. I was taking a hell of a chance as I struggled with this last group of surreptitious envelope openings. If I didn't pass the mandatory Flaps and Seals course, my overseas tour could be delayed or even canceled. But my instructor, Lynn, had specifically told us, "No French openings. We don't do those. Period, end of discussion."

A French opening meant slitting the end of the envelope, then gluing the paper back together. Yet there I was using the forbidden technique. If I messed up on this task, I'd be in serious trouble.

Lee, a research chemist from the Development and Engineering side of TSD and the other student in the course, was busy across the lab with his final problem, pretending not to watch me. But he knew I was countering Lynn's explicit orders, as I slit the *bottom* envelope fold and removed the contents, careful not to disturb any internal traps she might have laid for me.

I'd been able to conquer the other ten envelopes with relative ease, but I'd been told that the last five would be impossible to open "dry" with one of the tools we constructed out of wooden tongue depressors or the ivory from old piano keys. These last envelope flaps were also designed to prevent a "roll-out." This old Russian technique, using two knitting needles (or its Chinese variant with a split chopstick) was now considered too risky for intelligence work; it was almost impossible to slip the contents back into the envelope in the exact position where a sophisticated counterintelligence censor expected to see them. But I couldn't open the flaps either, because they were sealed with rice paste, impervious to mild heat and moisture. High heat and humidity would totally destroy the delicate Japanese paper.

I recalled leading kids into the dead-end box canyons outside Caliente to protect my precious bat guano caves, then climbing out the slippery sandstone chimneys in constant danger of a fatal fall. In the same spirit, opting for my own version of the forbidden French opening might indeed prove fatal to my career. We were supposed to just let these go, but I took a risk. I used a precise scalpel to cut through the bottom fold, reasoning that no one would expect this tactic. The typical method was to slice the much shorter side fold, which was much easier to repair.

Once I had photographed the contents of each envelope, I got down to the other step in my method—reweaving the paper fibers so the fold was completely restored and lacked a telltale glue line. This degree of hand-to-eye skill was not expected of students in the course, but Graphics artists aspired to meet these demanding standards on a daily basis. I now began using small amounts of adhesive to reconstruct the individual microscopic fibers along the cut area. This way the opening would not show even under ultraviolet light, which was used to reveal broken contrast patterns.

Looking up from the microscope, I was surprised to find Lee at my lab bench, grasping one of the restored envelopes.

"Amazing!" he said, thrusting the envelope under the bright light. "How in hell did you do that?"

Lee was taking the Flaps and Seals class to learn how to resolve some of the problems his branch faced by using enhanced technology. Lee had already seen me bring one innovation to the course by defeating the power of so-called "tamper-proof tape." The standard method involved delicately peeling back the tape with tweezers, using tiny drops of toxic solvent and a fine brush. But such prolonged exposure to poisonous fumes did not seem healthy or practical under stressful operational conditions. Therefore, when I tackled the exercise, I brought a can of rubber cement thinner and immersed the entire envelope with the tape into a glass lab tray filled with the solvent. The tape immediately floated to the surface, and after placing the tape on a clean glass sheet, I retrieved the envelope. The solvent quickly evaporated from all the materials, leaving them unblemished. I knew from past experience that this solvent would not stain or cause any ink to bleed. The success of this experiment sent Lee to his chemistry books.

"N hexane!" Lee had declared. "Bestine rubber cement thinner is N hexane." He had just scored a point for his branch.

"Necessity is the mother of invention," I had commented.

After completing my repair jobs, I packed up my envelopes and said good night to Lee, as he sat staring bleakly at his own remaining test envelopes on the workbench.

The next morning at nine sharp, Lynn examined the last fifteen envelopes in my course work. "Perfect," she said, shaking her head. "I don't know how you did it, but it's a great piece of work."

I sat back in smug silence. Now Lynn turned to Lee, bleary-eyed from his all-nighter in the lab. His first ten envelopes had only small defects, which she detected using glancing or ultraviolet light. She was following the standard routine of censors in counterintelligence services worldwide, whose job it was to verify the integrity of contents in their official pouches. In many countries, the postal services also employed security inspectors, who surreptitiously opened the suspicious mail of citizens suspected of espionage. In fact, the U.S. Postal Service had been created for that very purpose during the Civil War to serve as a central clearinghouse for mail censorship and counterintelligence against Confederate spies. During World War II, the Allies had a mail clearinghouse on the island of Bermuda. There, FBI and British MI5 counterintelligence censors, alerted by the agent Dusko Popov, first discovered German "microdot" secret writing, which contained densely packed, microscopic information. In fact, the use of earlier versions of microdots could be traced to the Franco-Prussian War of 1870. As with the historic practice of prostitution, totally new inventions in the espionage profession were rare.

Lynn shook one of Lee's final five tissue-thin envelopes, and the end fold split open. "What's this?" she asked suspiciously as Lee blanched. Then she seized my last five envelopes and thrust her hand inside. My meticulous weaving repairs came undone like zippers, and it was now my turn to cringe. These handmade paper envelopes were controls, meant to show us that there were some letters virtually impossible to open by standard techniques. In the field, if time permitted, the envelopes could be delivered to specialized labs for further processing, but attempting a French opening had been disastrous. Worse, my "innovation" demonstrated poor operational judgment.

"You may go back to your office," she ordered Lee. "I'll decide later how to deal with you."

She turned to face me with a menacing glare. "I ordered no French openings. You broke that order and are going to pay a price."

Fortunately, that price was not too high: six nights of cleaning the Training Branch labs in the South Building and repairing all of Lynn's equipment. One unexpected benefit of this penance was meeting with the legendary "Swift" brothers—two bachelor TSD engineers famous for their restless and inventive genius. They were pack rats of anything mechanical or electronic, truly devoted to providing the Agency with the most innovative spy gear imaginable. If anybody fulfilled the image of the Q Department "boffins" of the James Bond movies, it was the Swift boys. Although I got home late each night, the time I spent watching them at work was well worth the trouble.

Lynn kept me hanging for a week before announcing I would pass the course. It was only later that I came to understand what lay beneath her anger. I had violated a basic tenet in a profession where deception is the stock and trade: You never lie to or attempt to deceive a fellow officer in the service. Once you crossed this line, there was no going back. In what has been aptly termed a "wilderness of mirrors," a solid foundation of trust among colleagues had to exist.

I SPENT A total of twenty-one months at Headquarters. After I won my overseas assignment, I managed to jam an enormous amount of additional graphics and operational training into a brief period of time. In effect, I compressed several years of challenging specialty instruction into about six months—an exhausting experience.

When Karen, the kids, and I reached the futuristic concrete buttresses of Dulles International Airport on a cool afternoon in June 1967, prepared to depart for the Far East, I was tired but felt ready to confront the hostile forces out in that deeply troubled region. We would be living in military-style housing at a secret Agency base on Okinawa, supporting clandestine operations over a broad area.

As friends helped unload our suitcases from the station wagon, I realized the children were still too young to grasp the implications of our new adventure. But Karen's face bore a mixture of excitement and relief at having survived the chaos of medical exams and inoculations for the entire family, as well as the tedious routine of "packing out" and arranging our air-freight shipments, which would form the nucleus of our household until the larger surface shipment arrived. She had borne the brunt of the complicated logistics of the move; closing a house and relocating to Asia with three children under six was an enormous undertaking.

Somehow, we all managed to settle aboard the comfortable Northwest DC-8, en route for Tokyo and points south. I sensed that my family and I were entering strange, uncharted territory that would change us forever. I was right.

4 ■ Murky Waters, Southeast Asia

When you've got them by the balls, their hearts and
minds will follow.

—"Fred Graves"

Hong Kong, May 1968 ■ The Cathay Pacific Boeing 707 banked sharply
right, its gleaming wingtip pointing straight at a junk gliding across the
surface of the South China Sea. The two-masted junk was close-hauled,
tacking against the southern monsoon breeze from the mainland to
Hong Kong. I watched the setting sun gild the junk's sails like the wings
of a butterfly.

Embarking on my first Temporary Duty (TDY) away from the base
on Okinawa, I felt the mounting excitement of a fledgling voyager approaching the exotic world of Asia. Back at the base, my family and I
lived within the military cocoon: We shopped at the commissary and
PX. Milk was pasteurized, and the children ate fast food at the base

club. Karen and I played golf on weekends. My TDYs in the coming years would take me places where few of the locals had even heard of such amenities.

The plane completed the 180-degree turn on its approach toward Kai Tak International Airport, heading toward the spit of land which jutted from the Kowloon Peninsula into Hong Kong Harbor, one of the world's busiest deep-water ports. I caught a glimpse of a luxuriously green pyramid straight ahead—Victoria Peak, rising steeply behind the dragon's teeth of Hong Kong's skyscrapers. Lights were coming on, which made the dark vegetation an even richer green. The calm, gun-metal harbor teemed with thousands of ships and boats of all sizes, ranging from tiny sampans sculled by a single fisherman to the hulking gray monolith of an anchored American aircraft carrier, on leave after several weeks of air strikes against North Vietnam from Yankee Station in the Gulf of Tonkin.

I heard the grinding whir of the flaps extending, then the comforting thud of the landing gear locking into place. But suddenly, something seemed wrong. Instead of lining up straight on the final approach to-ward the string of twinkling strobe lights, sprinkled amid the neon glare of the tenements ahead, the plane banked hard right again while de-scending fast. *Oh, my God*, I thought, with an adrenaline rush, *they've lost control.* The jumbled television antennas, chicken coops, and elab-orate bamboo laundry poles on the flat roofs of the high-rise tenements seemed to reach up and grasp the right wingtip. In the airliner's blinking green running light, I could see the faces of the Chinese hanging their washing. A big Ralston Purina billboard loomed ahead.

Suddenly the plane snapped back, and the pilot eased the throttles. Crouching in my seat, I dared a glimpse at the narrow canyons between the tenements, the shadows where the routine bustling street life of an Asian city continued undisturbed with vendors peddling fried squid and scribes sitting in their underwear, slowly pecking at elaborate Chinese-character typewriters.

We're going to live after all. The plane skimmed over a wall of billboards plastered with garish ads, and I saw the runway's parallel amber lights. Then the plane slammed down on the concrete; the pilot applied full brakes and engine reversal, and we fishtailed twice before slowing to a controlled taxi.

On that first trip into Kai Tak, I was impressed by the straightforward and efficient nature of the British colony's immigration and customs controls. The officials manning the arrival lines were typically young Chinese men and women who had been trained to gaze intently, but nonaggressively, into the eyes of the traveler, betraying no emotion, but making it clear they were alert to deception. I did not encounter any of the confrontational behavior exhibited by arrival officials at third world airports. The officials at Kai Tak exuded pure business. Their midnight-blue uniforms were neatly tailored, trousers and skirts alike, with military, cable-knit sweaters, starched white shirts, immaculate ties, and shiny black shoes. They were a class act.

I observed these details not merely out of personal curiosity, but also because of professional interest. After a year of work and additional training at my Okinawa base, this TDY was to serve several purposes. One was to file a report on immigration and customs procedures in as much detail as I could possibly obtain without arousing suspicion. Espionage often involved moving people from country to country, and frequently these people had to travel under false identities, using altered or forged documents. My apprenticeship involved not only the preparation of these documents, but also the "building" of convincing cover legends for the illicit travelers, whether they were agents controlled by case officers or foreign defectors under hostile pursuit, seeking asylum in the United States. Sending me through an ostensibly friendly international crossroads such as Hong Kong was an excellent test of my ability to observe. At the end of this TDY, I would have to prepare both a "probe" report on the controls and an infiltration/exfiltration plan for Hong Kong to be used by defectors traveling in alias.

The female immigration officer flipped quickly through my burgundy official passport, inoculation record, and ongoing airline tickets, expertly absorbing the details.

"How long will you stay in Hong Kong?" Her polite question was spoken in perfect, clipped English. My answer was less important than the way in which I returned her unwavering gaze. Although she was a small person, her dark eyes felt large and powerful. I sensed strongly that the indignant-Western act would have little impact on these immigration inspectors.

Satisfied by my demeanor, she slammed the All Square Dater onto a blank page in my passport, depositing the blue-black impression of the arrival cachet. She initialed the lower left-hand corner of the Dater stamp with an indecipherable swirl. I noted that the Dater employed Roman numerals for the month, and I knew there was also a random rotation that included Arabic numerals, full words, and abbreviations for months. The employment of the sequence was a closely guarded secret. I was one of many TSD officers whose job it was to detect this arcane lore so that our duplication of passport cachets would be accurate and not trigger alarm bells among airport security personnel.

The officer handed back my papers with a perfunctory smile, but her eyes had already shifted to the next person in line. Carrying an official passport, dressed in PX sports clothes, I was just one of thousands of Americans working for the government in some capacity during the endless years of the Vietnam War. The fact that I did not represent the U.S. government agency I had claimed to when completing my disembarkation card was less important than my demeanor before that official. Impeccably forged or altered travel documents were only one-half of the equation in espionage travel: The ability to bluff convincingly while carrying them was just as crucial as "good docs."

Over the next few years, as my travel throughout the region increased, my worn passport overflowed with immigration cachets and visas, even expanding into eight-page foldout sections. That passport carried a rainbow collage of cachets: Wattay Airport in Vientiane, Saigon's Ton Son Nhut, Rangoon, Manila, and Bangkok. The efficient immigration officers at Kai Tak International slowly scrutinized each page. Perhaps they thought that I was traveling too much for a U.S. government rice production officer or administrative type. Just when I thought the immigration police were caricatures of the Inscrutable East, an officer stamped my passport and handed it across the counter, with her right thumb covering the first four letters of the word OFFICIAL and her left thumb obscuring the last letter. The gap left between her hands read: CIA. Neither of us spoke or smiled. I took my passport and walked toward the Nothing to Declare line of customs control. As I passed the end of the immigration counter, I noticed a thick, loose-leaf volume—the immigration police "Watch List." *Is my name in there?* I thought, wondering how secure these police files were from scrutiny by the opposition.

But that level of sophistication took me several years to achieve. On this first night in Hong Kong in May 1968, I fit the tourist persona more than the would-be street spook I'd pictured myself to be during exercises at the Farm. The brief taxi ride from Kai Tak swept me through the crowded streets of Kowloon, past the Peninsula Hotel, where a line of gleaming Rolls Royce sedans was arrayed, their liveried drivers tenderly stroking the lacquered bodywork with long feather dusters as they awaited the arrival of their passengers.

The taxi dropped me at the Star Ferry landing, and I was immediately engulfed in a churning crowd. I noted that my medium stature was hardly a drawback here. Traveling light, with just a canvas Pan Am carry-on bag and a large briefcase, I was able to slip through the human wave flowing toward the ferry ramp. It was a simple, preliminary Surveillance Detection Run. Any taller Westerner following me on foot would have been easily detected, moving diagonally through the crowd. I allowed several hundred people to clamber aboard the ferry, then jumped on myself just as the ramp was being lifted. Sailors in neat blue-and-white uniforms cast off the lines, the engines rumbled, and we churned out into the darkening harbor toward the sparkling lights of Hong Kong island.

Now my artist's eye overcame the emotions of a gawking tourist. The early monsoon had come, overpowering the dust and haze of the baking dry season. The sunset's afterglow above Victoria Peak created a halo of deep violet, dark crimson, and burnt orange. I snapped a picture with my Spotmatic. As we forged relentlessly ahead, Hong Kong's gleaming towers rose higher. We entered the main crossroads of the harbor, and the swirling wakes of hundreds of vessels reflected the dying sunlight in a kaleidoscope. The ferry was suddenly surrounded by boats and ships: freighters, other ferries, junks and sampans, poorly lit barges carrying loads of sand and rubble, and wallah wallah boats transporting whole families from the colony's outer islands.

Standing on the open deck with the humid tropical wind in my face, I wondered how the ferry captain managed to avoid the dozens of possible collisions with other vessels that crossed our track. Our air horn blasted, and the other vessels sounded urgent warnings in a variety of pitches. I had never seen or heard anything like this. It was Asia—crowded, busy, and chaotic, yet somehow organized along incomprehensible patterns.

For me, the sunset ferry ride was over much too quickly, and I found myself climbing the steep narrow streets between tier upon tier of Hong Kong skyscrapers, high-rise hotels, office blocks, and apartment buildings—some old and mildewed, others flaunting the gaudy steel-and-glass modernity of Manhattan or Los Angeles. Candy-pink and green neon lights replaced the glare of the sunset as I climbed the switchback roads of Central. Almost every block was broken by a strange Jack-and-the-Beanstalk latticework of bamboo scaffolding that vanished into the monsoon clouds, now colliding with the upper slopes of the mountain. In Southeast Asia, I would soon learn, strong and supple bamboo replaced the steel used for construction sites in the West.

When I checked into the Hong Kong Hilton on Queens Road Central, the officious desk clerk almost sneered as he asked if a bellhop could help with "the gentleman's suitcase."

"Airline misplaced it, I'm afraid," I said. "It'll be along tomorrow."

If all went well, I'd be out of this hotel and into a CIA safe house by the next night.

ACCORDING TO PLAN, the phone in my room rang around 8:15, and a dry English voice invited me to join him for a "whiskey and soda and a spot of dinner."

The invitation was the signal from the man I'll call "Jacob Jordan" to meet him at the rooftop bar at the Miramar Hotel on the Kowloon side. Jacob was already a mythical figure in the CIA when I began working with him in 1968. He was a senior TSD disguise and documents officer in the region, and undoubtedly had the most operational experience, moving both case officers and agents through difficult and dangerous areas. He also planned and conducted the exfiltration of defectors, often having to contend with hostile pursuit.

But the man I met for gin and tonic on the Miramar terrace that warm evening appeared to be anything but an American spy. Tanned and fit, with a distinguished military bearing, Jacob possessed British features and complexion, but his characteristically British qualities went far beyond appearances. That first night, he wore a tailored vested linen suit and a pair of expensive, custom-made Bond Street shoes. Everything about him, from his public school, regimental drawl, to the way he pinched his long Dunhill cigarettes between the thumb and index finger of his left hand, was evocative of a loyal British subject.

Yet I would soon learn that Jacob was in fact American-born and had been raised in a small town in the Midwest. Over the many years I worked with Jacob, often on long, exhausting operations, one would have expected him to drop his guard and reveal his true American persona. But he never did. Jacob was perhaps the best example I ever encountered of someone living his legend.

This permanent shift in style had been well justified. Jacob had entered the Army in the last year of World War II and was selected because of his aptitude for Oriental languages to attend the military language institute in Monterey, California. It was there that the kid from a prairie hick town completely lost his nasal Midwestern twang and was speaking Mandarin Chinese, Korean, and Japanese within three years. With the war over and his military obligation fulfilled, Jacob joined the CIA's old Technical Services staff, the predecessor of TSD.

His first overseas assignment was to the CIA base in Shanghai, just before the Communist takeover in 1949. Working with Chiang Kai-shek's Nationalist Kuomintang intelligence operatives, Jacob amassed a trove of documents and information about the security controls that Mao Tse-tung's Chinese Communists were implementing in the areas they had captured. By the time Mao's Red Army had taken control of the entire mainland, he was perhaps the leading American expert on Communist Chinese security and travel documents. He was then assigned to Japan, Taiwan, and, finally, the Okinawa base where I had my initial overseas experience.

While he was there, Jacob refined the basic document operations techniques employed by most intelligence services, adapting the American linear approach to security devices in official documents to more subtle Asian techniques. With his knowledge of the region increasing, he was able to detect patterns others could not. For example, a travel document might have been stamped with several hand-carved official "chops," bearing the approval of ascending levels of bureaucracy. In the West, the initial stamp of the most important authority would probably be the clearest and boldest. But on many Communist Chinese and North Korean documents, the chops of lesser officialdom were more prominent, while the obscure stamp of approval and initials of Party cadre appeared almost as an afterthought. Jacob's pioneering analytical eye and skill at replication techniques set the precedent for an entire generation of CIA disguise and documents officers, myself included.

Then the Korean War intervened, and.Jacob was posted to support irregular warfare operations against the Chinese and North Koreans. Next, the Agency sent him to Taiwan, where he helped authenticate Nationalist infiltration teams to be parachuted into the mainland by providing them with convincing documents and personal effects. Although these were bold and desperate operations, Jacob later explained, they were only marginally effective. As with similar efforts in Eastern Europe, the attempt to use OSS-type infiltrators and saboteurs in a Communist totalitarian state proved futile.

Nevertheless, his impressive language skills and ability to replicate almost any document, from a train ticket to a Party membership card, marked him as a "comer" with a brilliant career ahead. But this bright future suddenly came to a halt when Jacob fell in love with a beautiful mainland Christian Chinese refugee named Donna. Their marriage violated one of the Agency's most rigidly enforced policies: On marrying a foreign national, the officer must immediately resign. Headquarters tried to judge Jacob's case on its merits, but Donna still had an extended family living under Communist rule on the mainland, an unacceptable situation for an officer with access to sensitive operations. Jacob's security clearance was promptly downgraded to a "staff agent" contract, which meant he could no longer enter a secure CIA facility or read classified communications. He would be kept in this netherworld of espionage for several years until he arranged for his wife's family to leave the mainland and join them in Hong Kong, and was fully reinstated with TSD.

"Nice of you to drop by," he said, leading me to a wrought-iron table in the shadows at the far end of the terrace. I explained that my ferry ride back over had been "calm," meaning that I'd detected no surveillance.

"Lovely," Jacob quipped, signaling the waiter with a folded Hong Kong five-dollar note. "Let's just see if Victoria Harbor is still calm."

For the third time that night, I found myself on the Star Ferry, having completed yet another surveillance detection run up to the landing. Amid the confusion and hooting boat horns, Jacob explained our mission.

Chairman Mao's Great Proletarian Cultural Revolution was at its crest. Young Red Guards had decimated the ranks of the Communist bureaucracy, executing tens of thousands and exiling millions of intellectuals and other "parasites" from the cities to face starvation in the countryside. With the Vietnam War at its height, America desperately needed a better picture of the situation in the People's Republic of China

(PRC). Was the military still loyal to Mao? Was there a solid cadre of resistance within the Party around which anti-Mao forces might coalesce? These were crucial questions. Despite the destabilizing upheaval, China was still shipping thousands of tons of military materiel to North Vietnam each month and providing that "fraternal" Communist regime with many vital advisers.

Yet we had no diplomatic relations with the PRC, and therefore no embassy or consulates. Of course, we had no CIA station or bases on the mainland, nor did we have NOC ("Non-Official Cover") Agency officers permanently assigned to China, either in international organizations or under third-country business cover. In many ways, the vast nation of China, one-fifth of the world's population, was a hopeless enigma.

"We have to have agents in China," Jacob explained. "The PRC is one of U.S. Intelligence's major targets." But, he added, "most of the China watchers conduct their observations from here in Hong Kong."

As in Eastern Europe, he said, the CIA had assembled a system of legal travelers to serve as bridge agents. "But bridge agents are never really quite the thing, are they?" Jacob asked as the ferry nosed into the pilings on the Hong Kong side again. "With all the intelligence factories here in Hong Kong pumping bogus material, you can never tell when you're being led down the garden path." He turned to stare again at the mainland. "We need to bring important assets together for face-to-face meetings with our best case officers."

Making our way up the steep streets toward one of Jacob's favorite seafood restaurants in Wanchai, we mingled with throngs of GIs on R&R from Vietnam and sailors on liberty from the carrier task force in the harbor. It was easy to spot the differences among them. The real war zone grunts had what we used to call "ranchers' tans," their arms and faces burnt an umber brown, but their upper foreheads white from months of wearing steel-pot helmets. Most of the sailors were pale and overweight from long tours spent below deck, scarfing down ten-cent cheeseburgers from the ship's snack bars. The black GIs and sailors kept to themselves, a moody, vaguely threatening presence that served as a reminder that we were engaged in a geopolitical struggle in Asia while our own society still had unhealed wounds.

Moving along these narrow streets, Jacob continued to explain the nature of this particular operation. One of the most effective bridge agents had contacted an allegedly "top drawer" Chinese asset, a man with excellent Party and government connections who had undergone preliminary recruitment, now bore the cryptonym BARGER and was ready for an extensive meeting with a debriefing team in a nearby free port enclave. BARGER had legitimate business in the south of China and had already scheduled an extended holiday.

The situation was challenging. It was unlikely BARGER would be under internal security surveillance all the way from Beijing to the south. But without travel documents in an alias identity, BARGER could never pass in and out of the tight surveillance screen between the mainland and the free port. In addition, the combined allied debriefing team would have to be well disguised and documented in Crown Colony aliases, employing their most effective tradecraft because the free port itself was overrun with PRC counterintelligence officers.

The ethnic Asian bridge agent responsible for BARGER's preliminary recruitment had sent out his own travel documents, replete with legitimate entry and exit visas, to our office in Hong Kong via a smuggler's "rat line." Also contained in this sealed package were passport photographs and alias information for BARGER, who was due to meet our team in the free port within days. My job was to transform the bridge agent's travel papers into fully backstopped alias documents for BAR-GER, and I had less than twenty-four hours to do so. But that was not all. Once the meeting with the case officer debriefing team was completed and BARGER was safely back in Beijing, the bridge agent had to arrange to have the altered documents smuggled back to us in Hong Kong, where I would restore them to their original state.

The timing had to be precisely choreographed. Otherwise, someone could be badly "burnt," and in the PRC at that time, this often meant agonizing interrogation and execution.

Finishing my dinner with a clumsy display of chopstick work, I understood why Jacob had insisted we drink plenty of green tea at the end of the meal. We had a night of hard labor ahead of us.

"Got everything you need in your kit?" he asked, folding some Hong Kong dollars into the hand of the smiling waiter.

I patted my briefcase. Among my watercolor pads and brushes were the steel nibs, inks, glue, cutting and embossing tools, and passport binding thread I'd need for the job.

Jacob walked me to a taxi stand on the Admiralty waterfront and mumbled an address to a driver, noting he would meet me there in precisely thirty-five minutes. If he missed the first meeting, I was to return forty minutes later. He'd already briefed me on the need for such an elaborate countersurveillance "drill." It was always possible that either the trusted subagent or BARGER were doubles controlled by PRC counterintelligence. They might be trying to identify Agency officers and operations using Hong Kong as a base. Linking me with Jacob and an expanding circle of subagents would certainly earn some ambitious young PRC officer a promotion. But as Jacob taught me over the coming years, "Tradecraft is all in the details. If we can't accomplish the simple tasks, how can we take on the impossible missions?"

The safe house was a poorly air-conditioned, three-room apartment in a working-class Eurasian neighborhood. It had the advantage of both front and rear entrances and an exit from the rear courtyard onto two busy streets. A little Brit with a graying RAF sergeant's mustache soon appeared with the blank Crown Colony IDs for me to complete.

I spent the next six days and nights in that suffocating safe house, working under the hot glare of a watchmaker's lamp for hours at a time, or collapsing in complete exhaustion in the only air-conditioned bedroom, while the altered documents were being used during the two stages of the operation.

Once I had prepared the briefers' alias Hong Kong identity papers and delivered the package for Jacob to have smuggled back to BARGER on the mainland, I could relax until the documents were returned for me to work on them again. But Jacob insisted I stay indoors.

After five days, Jacob returned the altered documents, blithely commenting that the intricate debriefing operation in the free port had been a "really good show." BARGER had entered the enclave without mishap and conducted his extensive debriefing with the case officer team. They included a veteran polygrapher and an operational psychologist, who concluded the man was "solid." I felt more confident as I set to work reestablishing the bridge agent's identity on these well-used documents.

By the end of that week, I had begun to earn my spurs in real-world espionage, where people go down in flames if there are mistakes.

Much later, I learned that the information that BARGER provided had been of great value in the formation of our China policy. He was certain that Mao's Cultural Revolution did *not* represent the revolutionary future of China; instead, it was the final convulsion of internal class warfare between the Mandarin urban tradition inherited by the Party bureaucracy and the peasant populism exploited by Maoism. But BAR-

GER was not merely an abstract philosopher. He had provided essential practical details on the precise nature of the power struggle, identifying names and incidents.

Leaving Hong Kong on an overcast monsoon morning seven days after having arrived, I had no idea if the operation had been important to my government at all. But I had finally gotten a sense of the immense pressure I would experience as a field tech ops officer.

Little did I know.

Pakse, Laos, September 1968 ■ While most adult Americans know something about the war in Vietnam, few could explain why a CIA technical operations officer like myself would be strapped into the canvas sling seat of a World War II–vintage C-46 one day in September 1968. We were flying south from Vientiane, the diplomatic capital of the land-locked Kingdom of Laos, to the provincial town of Pakse on the banks of the Mekong in the south. The side door had been removed for rice drops, and the aisle was jammed with bamboo cages of chickens and ducks.

As a fragment of France's lost Asian empire, Laos had become another pawn on the Cold War chess board. When Ho Chi Minh's Communist Viet Minh forces defeated the French Army at Dien Bien Phu in 1954, the first Geneva peace conference on Indochina later that year officially divided Vietnam into two zones, north and south, but did not clearly establish the status of Laos and Cambodia. Laos became a constitutional monarchy, and I suppose the statesmen in Geneva convinced themselves it would somehow evolve into a neutral mini-Sweden in the cloud forests of the Annamite Cordillera (what the Vietnamese call the *Truong Son*, or Long Mountains). Instead, the country quickly slipped into a heated civil war, initiated by the Viet Minh–led Pathet Lao, who had been generously supported by the Soviet Union. The Communists soon drove the inexperienced and ill-equipped Royal Lao Army from the north of the country, and a series of coups and countercoups ensued, leading to a new Geneva treaty that divided Laos into rightist, neutralist, and leftist factions, backed by both the U.S. and the Pathet Lao, a Hanoi-Soviet surrogate.

Those who suffered most were the estimated three million members of mountain tribes, whom we came to call by their rightful name Hmong ("the People"), but whom the lowland Lao and the Vietnamese disparaged as the Meo or the *moi*, meaning "savage." After the 1962 accord guaranteeing the "neutrality" of Laos, both the U.S. and the Soviets were obliged to remove their military advisers. The U.S. Green Berets, who had trained and sometimes fought beside the Royal Lao Army, were withdrawn. But the Soviets retained their adviser mission to the NVA and the Pathet Lao, trying to operate more discreetly from airfields and depots around Hanoi.

The year I arrived in Indochina, the North Vietnamese had thousands of road builders, truck mechanics, and antiaircraft crews stationed along a vast web of rain forest roads known as the Ho Chi Minh Trail. Every year from 1967 onward, Hanoi had moved another 100,000 NVA replacements down the Trail and into South Vietnam. Yet political constraints still prevented the Johnson administration from openly breaching the 1962 Geneva Convention.

Therefore, one of the few ways America could respond to Hanoi's flagrant violation of the Convention was through a combination of bombing and covert military and CIA action. The U.S. Military Assistance Command, Vietnam (MACV) in Saigon ran cross-border reconnaissance, sabotage, and prisoner-snatch missions as part of its highly classified Studies and Observations Group (SOG), sending teams of local Nung and Montagnard troops, led by American Green Berets, into Laos from secret launch bases in South Vietnam. But the SOG cross-border teams were restricted to a narrow band of operational boxes close to the mountainous border. One of the SOG team's key contributions was to identify "lucrative" targets for airstrikes, such as ammunition caches, truck repair facilities, and the bivouacs of large units moving south.

The North Vietnamese responded by expanding the intricate network of high-speed foot paths and vehicle roads of the Ho Chi Minh Trail system to the west, where the CIA covert action took over. The Agency stationed road-watch and "striker" teams—composed of tribal irregulars and tough Thai mercenaries led by CIA case officers—on Lima ("Landing") Site bases, carved into the imposing limestone domes and monoliths above the Trail network and concealed by the unbroken triple canopy of the rain forest below.

I was flying to Pakse that rainy September morning, en route to Lima Site 38, one of the secret bases closest to the main axis of the Ho Chi Minh Trail in southern Laos.

My first impression of the local "knuckle draggers," as the paramilitary covert action boys were known, was mixed. Some looked like overgrown Boy Scouts, while others looked like paunchy retreads from several wars back. I was wrong on both counts. In my opinion, the case officers serving with the Hmong army and other anti-Communist tribesmen in Laos were in fact some of the best guerrilla leaders America ever put onto a battlefield. Almost all of them were well-trained military veterans, most with Special Forces combat tours in Vietnam behind them, and many with experience as advisers to the Army of the Republic of Vietnam (ARVN) under their belts.

One of these men, whom I will call Tommy Lobo, married a Lao princess. I'd already read about his exploits in *True* magazine when I met the real man on his remote Lima Site fortress in northern Laos. I had heard the tale about Tommy concerning an argument he'd had with Headquarters over his NVA "kill reports," which some bean counter claimed he had inflated to raise the status of his Hmong strikers. In search of evidence Tommy instructed his men to bring back the severed left ear of each kill they had made. He then wrapped the ears and sent them in his next sealed dispatch to CIA Headquarters. The Headquarters desk officer who opened the pouch never again discredited one of Tommy's postaction reports.

MY JOB AS a tech ops officer and an artist in the regional Graphics unit was to work with these teams on a wide variety of propaganda and psychological warfare operations. I knew volunteering for this mission would take me to some exotic places, but I wasn't quite prepared for the scene I encountered my third morning in Pakse, when Buck Foster rose early and rousted me from my sweaty cot with a mug of coffee.

"It's still blowing a typhoon," I complained over the drumming beat of the rain on the roof.

"Just a Mekong drizzle," Buck responded.

At the airfield, the clouds hung low and heavy in the east, but Buck seemed to have reached the conclusion that flying up to Lima Site 38 was both possible and prudent. I stood under the roof of the air ops building, staring dubiously as the ground crew loaded the plane that would take us there. It was a mud-spattered Swiss Pilatus Porter that looked like an overgrown Piper Cub with its single, Pinocchio-snout, turbo-prop engine. But this Short Takeoff and Landing (STOL) aircraft was highly regarded as a reliable Agency bush taxi and "trash hauler," capable of delivering passengers and enormous loads to remote postage-stamp airstrips.

I certainly hoped the plane's renown was well earned, as I watched the Thai ground crew heave wooden ammunition crates, bulging white USAID rice sacks, and stacks of weapons rolled like crude cigars in ponchos through the plane's single cargo door. Once this load was more or less lashed down with webbing, Buck turned to the small crowd of tribal irregulars and Thai mercenaries lounging around the building.

"Time to board," Buck said. "We've got to get you assholes to work."

I assumed two or three of us were going to clamber aboard the plane and find places among the sacks and boxes. Instead, a total of fourteen men, albeit some of them very small Kha soldiers, dashed through the rain and squeezed into the narrow confines of the cargo hold. With no seats we had to stoop with our heads and shoulders hunched against the lightly padded overhead. The compartment opened forward, revealing a two-seat cockpit with wide Plexiglas windshields.

The pilot, a dapper Thai in neatly pressed khaki uniform replete with crossed bandoleers and two pearl-handled .45 revolvers, picked his way among the puddles under the shelter of a wide umbrella held by a barefoot lackey. His audacious outfit crystallized the impression that had been growing for several days that somehow I had slipped into a time warp back to the U.S.–Mexican border skirmishes between Pancho Villa and General Black Jack Pershing, that the Mekong was actually the Rio Grande and Vientiane was Dodge City.

But the stout little dandy in the cockpit knew what he was doing. In a blur of deft movements, he started the turbine engine, throttled it up to a nose-itching howl, released the brakes, and taxied smartly to the end of the strip. Then we were slamming forward, the turbo-prop screaming as we bumped over the rough steel-mat runway. Only seconds later, the plane bounced from the ground and was sucked into the roiling clouds. Climbing, the engine throttled to the max and the prop shrieking, the stuffy cargo hold was constantly buffeted. A skinny young Kha trooper was wedged onto rice sacks to my left, clinging to a tie-down strap clipped to an aluminum rib. *If he pukes now,* I thought, *there's no place I can hide.*

Just when it seemed that the engine would rip itself apart, the pilot throttled back, and we seemed to stop, suspended in a jar of hot milk. The sensation was terrifying. I *knew* we were falling, yet had no sense of the roller-coaster weightlessness I should have felt. I'd been seized by vertigo, which seemed to be endless, but I didn't want to look at my watch and see we were overdue on the scheduled forty-minute "hop." *There's no way this overloaded plane can still be flying.*

Then we popped out of the overcast into the glare of a tropical morning sky. Through the side window I saw we were climbing smoothly away from the wispy tops of the cloud deck. But looking ahead, I felt a stab of adrenaline. We were flying straight toward a vertical, gray limestone wall, which disappeared into a shroud of mist a hundred feet higher than our flight path. A beautiful creamy waterfall arched from a mist-hidden cliff and disappeared into clouds below. That morning, I'd studied the air force navigation chart on the wall of Buck's office, noting several ominous near-vertical humps blandly marked "karst." These were jungle-covered monoliths eroded into free-standing towers, like so many unlit phone booths scattered along a dark highway.

The pilot rammed the throttle forward, the engine screamed and whistled, and the overcrowded Porter seemed to rise straight up again, into the mist lapping down from the stony mountaintop ahead. *When will we hit?* I was trying to remember an appropriate prayer, while simultaneously wondering why I had abandoned a perfectly safe and honorable job in the Graphics bullpen in Washington.

Then the mist shredded, and I could see the craggy top of the karst spreading into a flat mesa covered with mixed hardwood rain forest and scrub jungle. The pilot banked sharply right, lining up with the muddy orange scar of the landing strip that appeared suddenly from a dense forest grove. In another blur of moving hands, the little Thai deployed wide wing flaps and slats, throttled back, and raised the nose toward the vertical. As I caught sight of some bamboo huts and a sandbagged bunker beside the strip, the Porter touched down with amazing lightness on the rutted laterite mud and stopped within forty feet, more like a winged helicopter than a conventional airplane.

I had arrived at Site 38, perched one thousand feet above NVA Supply Route 92 on the Ho Chi Minh Trail.

Before I could absorb my surroundings, the Kha soldiers had dragged the cargo from the hold, and the Thai pilot revved the Porter's engine for a quick takeoff. He jolted about thirty yards across the ruts, then lifted off like a dragonfly and disappeared once more into the clouds.

"Welcome to Dogpatch." The resident Agency case officer in command of the Site's strikers offered me his hand. A clean-cut, former Special Forces captain, he looked about my age, but had already served over five years in the Indochina war zones. He was one of the many paramilitary men I would work with in the coming years, some of whom I would come to know on a close personal basis. Case officers were supposed to send their local troops into battle and stay out of harm's way themselves, but most ignored this policy. According to official accounts, only five CIA officers had been killed in Laos by 1973, the end of America's paramilitary involvement in Indochina. The accuracy of those casualty figures is definitely open for debate.

Like many of his colleagues, the officer in charge of LS 38 preferred a neutral "handle" (in his case, "Ridge Runner") because he spent so much of his operational life at risk of capture and did not want Hanoi to exploit him as a CIA spy in command of foreign mercenaries. Some covert action case officers in the more exposed Plain of Jars went to war wearing T-shirts and Levi's. Ridge Runner, on the other hand, favored a composite uniform more appropriate for the jungle trails of the cloud-choked valleys below. He wore the black-and-green, tiger-striped camouflage ARVN Ranger shirt and olive-green, cargo-pocket GI trousers, cinched tight with rubberband leech straps at the ankles. Instead of the familiar American jungle boot with its distinctive cleated sole, Ridge Runner sported the standard ankle-high, canvas Bata boots worn by many NVA units. "Even if you're good in the woods, everybody leaves footprints," he later explained, "and I don't want my people ambushed because some NVA tracker picks up the pattern of my boot soles in the mud."

Ridge Runner's attention to detail made my work easier. Whenever his team pulled off an ambush or snatched a prisoner, Ridge Runner studied the papers retrieved from the enemy, personally interrogating the prisoner before turning the "take" over to his superiors. He knew both the NVA and Pathet Lao order of battle in his region intimately, as well as the shifting patterns of enemy morale.

Ridge Runner also proved to be an excellent resource for my office's propaganda and "psy-war" (psychological-warfare) efforts. Although the 1968 Tet Offensive had been a political and psychological victory for the Vietnamese Communists, the massive assault against the towns and cities of South Vietnam had been a military debacle for their cause, resulting in horrible casualties. Our reconnaissance had revealed that tens of thousands of NVA troops were back in their sanctuaries inside Laos and Cambodia, "licking their wounds." Ridge Runner had urged us to intensify our surrender pass campaign, tailoring the message to reach weary, malnourished NVA troops, forced to sleep in half-flooded bunkers. Carved into the forest mountainsides along the five-hundred-mile section of the Trail bordering South Vietnam, these ramshackle structures were under constant threat of being bombed by high-flying B-52s.

Ridge Runner and his colleagues helped me design illustrated propaganda similar in format to colorful comic books. Intended to impress NVA troops of peasant origin and the poorly educated Pathet Lao, my drawings showed how the Communists had overrun a peaceful land, disrupting the simple life of the tribal people in the mountains, stealing their crops, and enslaving them as porters on the Trail, where many were killed in air attacks. We interwove an underlying theme of karma and redemption into these visual messages, playing on Buddhist beliefs which went far deeper into the hearts of the people than did Communism. Our message was subtle but clear: Unjustly cause pain and death in this lifetime, and the wheel of karma will turn on you and your family.

Over the next five years of American military involvement in Indochina, several thousand Communist troops in South Vietnam and Laos would rally to our side. Most bore our surrender passes, while many others cited the corruption and hypocrisy of their leaders in North Vietnam as compelling reasons for their surrender.

Vientiane, Laos, November 1968 ■ The only light in the bungalow's dining room came from the open window behind me, perfect for the delicate task at hand. The morning was overcast, and the typical tropical glare of the sun was filtered by thick clouds. Coming over my left shoulder, the diffuse daylight gave me the steady illumination I needed, free of distracting shadows.

Hunched over the dining table, I was almost paralyzed with concentration; the muscles in my neck, shoulders, and right arm were in painful knots, caused by the fierce effort to maintain control of my fingers on the fountain pen. My pen was held steady by a taut, two-hand "forger's bridge" formed by my left and right arms—elbows locked against the table, wrists arched, with the fingers of my left hand intertwined with those of my right.

The young woman across the table watched me silently. She sat frozen in place, holding her breath as I followed the proper stroke order of the intricate Asian characters I was inscribing on the blue-lined page of the narrow diary. The characters were totally incomprehensible to me; fortunately, the Asian-American woman with me was the wife of an Agency NOC officer and had been schooled in the language as a child.

Although her presence was one of the positive points of this operation, it was far outweighed by the risks we were taking. If this entry was not exactly right, or if my fountain pen erupted in blots, as it was prone to do in tropical humidity, we would have to re-create the diary, with its complete set of handwritten entries, in the desperately short time available.

But the ink was flowing perfectly after I put a single drop of gum arabic in the pot and refilled the bladder. I applied and released pen pressure on the paper in a slow, rocking motion made possible by the forger's bridge. This technique allowed me to duplicate the writing speed and pressure of the diary's owner, whose swirling Asian characters I was imitating. A single drop of sweat ran down my ribs into the waistband of my trousers, and my locked fingers were throbbing painfully as I finished copying the remaining passages in the same casual style we had practiced several times that morning.

Her somber gaze was the only obvious sign of her role in this operation. She wore a plain housedress, and her long, straight black hair was tied back with string. My forgery materials were spread out between us on the teak tabletop. Two well-behaved toddlers were playing with matchbox cars around her bare feet on the faded ceramic tile floor. They chattered in a strange mixture of two languages, sometimes piping up loudly. But their mother and I were completely immersed in the complicated task at hand, working in the high-ceilinged dining room of this shabby villa on the outskirts of Vientiane.

I heard motorbikes on the dusty road outside, then the rumble of military trucks in a convoy approaching from the airport. "If they kick up much dust," I said, keeping my voice steady, "I'll have to ask you to close the windows."

Looking up quickly, I saw her nod in understanding. My eyes moved to the lined pad lying near my right elbow, which contained the template of entries we had worked on for several hours. I had two more to finish, and the job would be done. As usual, I was operating with a sleep deficit. I'd arrived from Bangkok early that morning aboard a CIA contract flight on an old C-47 heading north to the air base at Udorn, Thailand, and then jumped onto a Porter across the Mekong border to the Vientiane airport. It was one of those no-passport-stamp trips we risked on urgent operations. In this case, the urgency had been clear as soon as the Secret IMMEDIATE NIACT cable had hit my Okinawa base thirty-six hours earlier, requesting the services of AN ARTIST/VALIDATOR [FORGER] ON-SITE IN VIENTIANE SOONEST.

As Mark, the case officer who had sent the cable, drove me from the airport to this villa along a circuitous route, he briefed me on the urgent job I faced.

"We've got one day—max—to complete some illicit alterations to papers and diaries," he said, then added details of what must have been one of the most convoluted operations of the endless war in Southeast Asia.

Several weeks earlier, the local CIA station had received word that two members of an Asian Communist party would enter Vientiane en route to Hanoi to help prepare "evidence" of American atrocities and war crimes. They were scheduled to present this propaganda at the dubious international war crimes trial in Stockholm, convened by Lord Bertrand Russell, the elderly British pacifist well known for his naive view of Communist intentions in the world. The station had discovered that their travel itinerary included arrival and departure stops in Bangkok and Vientiane.

The Royal Thai and Royal Lao governments were our allies in suppressing the Communist insurgencies in their countries, but none of our liaison contacts among the local customs or immigration officials felt they could deny the Asian Communists passage through their airports en route to Hanoi. Upon arriving in Vientiane two weeks earlier, the two men had been placed in a modest hotel near the river, courtesy of the North Vietnamese embassy, while they waited for the Aeroflot plane to Hanoi the next day.

Since the Station had orders to deflect them from their mission and the local authorities would not cooperate, Mark and his fellow case officers took matters into their own hands. When the Communists left their hotel for dinner that evening the local TOPS officer helped the case officers enter their rooms and plant what he called "compromising materials" in the men's luggage. He hoped Lao Customs at the airport would find the incriminating documents in the morning and deplane the men, cancel their transit visas, and send them back to Bangkok.

Unfortunately the Lao officials were too polite to ask the men to open their bags, and they departed safely for Hanoi.

The two Communists returned ten days later and were booked on a flight to Bangkok the next afternoon. The challenge now facing the Station was to capture the questionable evidence they would take to Stockholm. The Station had already alerted a small team of local young people known as the Flying Squad to conduct surveillance on them and snatch the evidence, reporting that the two men always carried a denim shoulder bag. Sure enough, a quick but thorough search of their room revealed that the evidence must have been in the bag, so Mark ordered the Flying Squad to seize it. That evening, as the Communists strolled down to a riverfront café, two Squad members coasted by on a Honda motorbike. One of them jumped off, grabbed the bag, and leaped back onto the bike. They roared away, and had almost turned a corner when one of the Communists shouted, "Stop, thief!"

A Royal Lao Army lieutenant came to the rescue, knocking down the Flying Squad members, and retrieving the bag, which he proudly returned to the grateful Communists. Our boys escaped in the confusion, but the Communists still had their precious evidence.

Mark had a fallback plan. He managed to cancel the Communists' airline reservations and purchase every seat on all the commercial flights leaving Vientiane for Bangkok in the next week. But this proved futile as well. The North Vietnamese embassy then arranged for the Communists to take the ferry across the Mekong to Nong Khai and travel on the overnight train to Bangkok. Resourceful as ever, Mark turned to the next contingency plan, buying train tickets for the Flying Squad members in a last-ditch effort to snatch the bag. Simultaneously, the Station implored its Thai contacts to have their customs authorities confiscate the two Communists' luggage long enough for us to examine it. Our officers explained that the men were Communist spies conducting espionage operations in Thailand, and we would provide solid evidence. Our claim seemed convincing to the Thais, who were embroiled in a violent Communist insurgency in their northeastern hinterland. They promptly placed the two men under house arrest and allowed Mark's team to search their luggage, giving us only twenty-four hours out of fear of retaliation from the men's government.

What Mark discovered made everything worthwhile. The materials included a can of 16mm filmed statements from captured American pilots. These prisoners had been tortured into making false confessions of school and hospital bombings, as well as other grotesque offenses. There were also carefully prepared documents "proving" that the United States was engaged in "genocide" against Socialist Vietnam, while using defoliants and other forms of chemical warfare to destroy farms in the Red River Valley.

It was at this point that Mark decided to send the IMMEDIATE cable to my base.

"I THINK HERE he is drunk," the young woman said, pointing delicately at the diary page. "See how the characters become twisted."

I rocked back in my chair, pivoting on my elbows, and eased the tension of the forger's bridge. I capped my fountain pen and laid it aside to knead the muscles of my neck. Even though I couldn't read the words, I saw at once that she had detected a subtle point we had both missed during the long rehearsals that morning as we prepared our entries. Now we had to create a fresh template for these final two valuable passages, the most difficult forgery conceivable. Not only would I be adding entries to the same page in foreign characters, but I would also be imitating the idiosyncratic nuances of the writer, which varied from day to day. I sighed, exhausted and anxious, yet all too aware of the fact that I couldn't just hurl the pen across the room and relax with a cold beer. After a short break, we set to work again.

It was mid-afternoon before we finished. The entries had been brief and oblique, referring to numbered, unnamed documents. Our next task was to create very small notes in the same script on tissue-thin, water-soluble paper that were explicitly related to espionage: agent contact instructions, code names, and lists of clandestine networks in Thai cities and villages. We rolled these small notes into tight bundles and concealed them in the spine of the diary and throughout the confiscated material.

By the end of that long afternoon, I was totally spent but enormously impressed with the intelligence and sophistication of the young woman who had assisted me on such short notice. Only she could have provided the authentic phrasing and nuances needed to convince the world that the two Communists were, in fact, spymasters. We were finally ready for Mark to return the material to the Thais and wait for their reaction.

A MONTH LATER, I was sitting by the pool at the Imperial Hotel in Bangkok, leafing through the Sunday edition of the *Bangkok World*. A headline caught my eye: NEST OF SPIES UNCOVERED IN NORTHEAST THAILAND. The article went on for several pages, explaining how two members of an Asian Communist delegation entering Thailand from North Vietnam had been arrested for carrying illicit materials indicating that they were involved in espionage against Thailand. A full translation of all the diary entries and the concealed secret notes discovered in the confiscated material followed. This suggested that "some of the evidence being held by the Thai Government also appears to be part of a Communist operation to spread lies about the United States . . ." The article concluded by noting that the Communist party in the two delegates' homeland had protested the "false charges" against their members and demanded that the compromising materials be examined by an independent forensic expert in their capital. The Thais had agreed, and the expert had declared the materials to be genuine, not forgeries, as the Communist party had insisted. Satisfied by the outcome, the Thais had released the two Communists but retained all the confiscated material, which was never to be released for public consumption.

In retrospect, was this just another meaningless CIA "dirty trick," a desperate attempt to prolong a lost war? Not by a long shot. If North Vietnamese military intelligence in Hanoi had the sense that they could get away with the torture of American POWs and peddle valuable propaganda films at international forums, the merciless atrocities committed at notorious camps such as the Zoo in Hanoi, orchestrated by a brutal Cuban interrogator the Americans called "Fidel," would have continued indefinitely. More Americans would have died under the whip and the bamboo bludgeon. In fact, the escalating campaign to extract propaganda from American POWs in Hanoi ended the next year. Did the success of the diary operation have something to do with improving conditions? We will probably never know.

But I do know with certainty that the North Vietnamese Communists were in no position to launch an international propaganda offensive against American war crimes when their own hands were so bloody. The Viet Cong and NVA routinely tortured and executed South Vietnamese government officials, all the way down to the village school teacher, in an organized effort to exert control over rural areas through terror tactics. During the NVA occupation of the old imperial capital of Hue in the 1968 Tet Offensive, Communist execution squads massacred as many as three thousand unarmed prisoners, including "cruel tyrants and reactionary elements," who had been slated for extermination months earlier. Commenting on the worst massacre of the Vietnam War, Pulitzer Prize–winning historian Stanley Karnow noted that the victims "had been shot or clubbed to death, or buried alive." He added that America had hardly noticed those atrocities because it was preoccupied with "the incident at My Lai—in which American soldiers had massacred a hundred Vietnamese peasants, women and children among them."

Dalat, South Vietnam, October 1969 ■ "Hang on to your lunch," the contract pilot said with a wry grin, dropping the nose of the Beech Twin Baron toward the rolling green landscape below. We dived in a tight corkscrew, trying to stay within the invisible protective cone of ARVN control, which ended at the perimeter of the small airstrip below.

This trip to the lovely old colonial town of Dalat in the "Switzerland of Vietnam" was typical of most of my assignments during six years in the war zone. As had been the case in Laos, I had become one of the leading artist/validators supporting Agency operations in Vietnam. But my artistic abilities were developing beyond the rigid discipline required in the straightforward duplication of enemy documents. In the next several days, my skills would be rigorously tested.

The plane jolted onto the muddy strip, and the pilot immediately jammed on the brakes. I soon found myself standing in the cool sunshine of the Central Highlands, watching the Baron roar down the runway and spiral into a tight, reverse corkscrew climb. Besides the ARVN troops manning sandbagged pillboxes on the perimeter, I was all alone.

The Viet Cong had hit the local marketplace two days before in a grenade attack, killing and maiming innocent people to discredit the South Vietnamese government and to demonstrate their ability to penetrate a tightly guarded provincial capital. I felt the familiar surge of adrenaline and anxiety, which always seemed to haunt me before a new mission.

Suddenly, I heard the roar of an engine behind me. An elegant 1939 Citroën *traction-avante* lunged into view like a cougar, lurched across the field, and skidded to a stop beside me. Two Americans in unmarked fatigues, carrying CAR-15 assault rifles and draped in bandoleers bulging with extra clips, jumped out to scan the runway.

"Sorry to be late," one of the case officers said, grabbing my duffel and throwing it into the cavernous rear seat of the Citroën. The other American shoved me in after my bag. In a flash, they were back in the front seat, and the driver threw the car in gear and stamped on the accelerator. Both men kept their short-barreled weapons pointed out the open windows, anticipating an ambush. We were careening down the switchback dirt road from the airport to town before I realized that this ancient saloon car actually had an eight-track stereo. As we bounced along at reckless speed, Johnny Cash was singing "I Walk the Line." The scene was so bizarre that I exploded in fits of laughter as I tumbled around the backseat.

The local CIA base consisted of an old colonial French villa with a sandbagged gate and windows screened with cyclone fencing to thwart Chinese-made B-40 rocket-propelled grenades. The base also operated several safe houses near the ARVN military academy. It was there that I met one of the most intriguing people I would encounter during my trips in and out of Vietnam.

A woman I will call Ming, in her early twenties, was a former member of the Viet Cong. She had rallied to the government side under the *Chieu Hoi* program. In itself, this action was not unusual as the war dragged on, but her story was. For several years Ming had been the cook at a Communist safe house used as a way station for infiltrating North Vietnamese *Trinh Sat* intelligence service officers off the southern terminus of the Ho Chi Minh Trail and into Saigon. Their dual mission was to penetrate the South Vietnamese government and ARVN, and to establish clandestine communication links with Hanoi. When Ming revealed this, we had to find a way to identify these key intelligence operatives, because it was obvious that their presence in the South was compromising the effectiveness of Saigon's war effort.

But Ming was an uneducated peasant girl and, being good operatives, the Hanoi officers never used true names or revealed personal details in her presence. Ming, however, had a near-photographic memory. I had already helped South Vietnamese counterparts prepare police sketches of suspected VC cadres by debriefing witnesses, but the trail Ming and I had to follow stretched back years. Initially, I wasn't optimistic. Then, I spent several hours talking to Ming through an interpreter and realized she was indeed a remarkable source.

It turned out that Ming had a romantic streak, a quality discouraged by the Viet Cong. To amuse herself, she had created fantasy tales for each of the several dozen people who had passed through the safe house en route to Saigon over the years. These tales were her mental cues for recalling their exact appearance and mannerisms.

Working with the interpreter, Ming would patiently describe the fantasy image she had created for each real person. Then we would study albums with hundreds of photographs of Northern and Southern Vietnamese, and she would select certain features shared by the subject of her fantasy and the people in the photographs. As I sketched, combining visual details from the photographs with her rich descriptions, Ming would suggest refinements. For example, one of her stories concerned the "worried student" who had lost his books and developed the nervous tic of twisting his left earlobe. Another character was the "impatient doctor" who cleared his throat and brusquely interrupted people before they could complete their statements.

After two and a half days of debriefings and sketches, Ming and I had finished twenty-six face-on and profile portraits of important Communist intelligence infiltrators. Although it occurred to me that she could have been nothing more than an excellent storyteller with a vivid imagination, I did not think so by the end—her descriptions were simply too detailed and consistent.

As events turned out, South Vietnamese counterintelligence officers made thirteen arrests in the following months based on those portraits. Each of the suspects confessed, and most were caught either red-handed, engaged in acts of espionage, or carrying spy paraphernalia. They were all involved in running local agent networks, consisting of Viet Cong who had penetrated the foreign community in Saigon. Their agents were our trusted servants and employees, supposedly "vetted" by our Vietnamese counterparts. For years, these operatives had enjoyed privileged access to the homes of Americans working in Saigon.

Savannakhet, Laos, July 1972 ■ American ground units had withdrawn from Vietnam earlier that year, leaving behind small detachments to guard coastal enclaves such as Danang and Saigon. But the devastating use of American air power three months earlier to defeat the NVA's massive Easter Offensive revealed that military issues still needed to be resolved. Nevertheless, America wanted out of the war, which threatened to destroy our society from within and render us impotent on the larger geopolitical stage of the Cold War.

In July 1972, with the American pullout almost complete, the Agency still had important responsibilities in Southeast Asia. Richard Nixon's National Security Adviser, Henry Kissinger, was entangled in frustrating, secret peace negotiations with the North Vietnamese in Paris. The issue of American Prisoners of War, especially those known to have been captured alive in Laos, was one of the most contentious points of discussion. Even in the secret talks, the Hanoi negotiators steadfastly refused to admit they had troops in Laos, or that they controlled the Pathet Lao. The military and the CIA estimated, however, that there might have been more than fifty American military and civilian air crew members held captive in caves near the Pathet Lao capital of Sam Neua, in the northeastern corner of the country. It was vital for

Kissinger to confront the North Vietnamese negotiators with facts, not estimates, and, unlike Lyndon Johnson and Richard Nixon, who often expressed contempt for "those clowns over in Langley," Kissinger depended on the Agency for fast and accurate intelligence.

Kissinger's needs in Paris had brought me to this seedy market town on the Mekong in southern Laos. I had already helped develop clandestine channels of communication to and from captured pilots. We analyzed their handwriting in the occasional letters to their families, which the North Vietnamese allowed, and realized that some of the POWs were placing minute, innocuous dots around certain letters in a code. TSD also maintained a collection of more than fifteen thousand photographs, culled from various enemy military, Chinese, Soviet bloc, and European press sources, which might have shed light on missing American servicemen. Each of these photos had been carefully analyzed and rated on a scale of authenticity so that our peace talk negotiators would have a realistic idea of how many Americans might have reached enemy captivity alive.

My job in Savannakhet related to this issue involved both basic and more sophisticated techniques. The CIA command in Indochina was preparing to dispatch an "indig" team of local agents, composed of *Chieu Hoi* defectors from the NVA and Pathet Lao, up the Ho Chi Minh Trail to the enemy headquarters at Sam Neua, in a desperate attempt to verify how many American POWs were held there. Kissinger could make good use of any such hard intelligence in Paris, so the potential advantage outweighed the risk. This type of operation was known in the trade as a "10 percenter," meaning that there was a 90 percent chance of failure.

The man on whom the fate of the operation largely depended was a wiry, former North Vietnamese officer in his early thirties, who sat across a table from me in an Agency hooch, watching intently as I worked on an array of NVA documents. He had been a senior lieutenant in the NVA's General Political Directorate (*Tong Cuc Chinh Tri*), the important political commissars who played a variety of roles, ranging from secular chaplain to co-commander of combat formations. His particular unit, a crack Hanoi Guards regiment, had been ripped apart by American B-52 Arc Light strikes hitting Rocket Ridge above Kontum during the NVA's ill-fated Easter Offensive. Our counterintelligence estimated the man was a "50-50," meaning that he presented even odds of sincere defection. But sending him back among the NVA and north along the Trail on this mission was worth the risk of his possible redefection.

He had advised us that a man in his position would travel the gravel roadways of the western Trail, accompanied by at least two Pathet Lao or NVA-enlisted bodyguards. We located two suitable defectors to form his team, and all his equipment was assembled by Saigon's TSD operation, working with MACV/SOG. Their vehicle would be a GAZ 69 A series four-wheel drive, Soviet-built jeep with a detachable canvas roof, bearing serial numbers and identification decals of the NVA's 559th Transportation Group, which managed Trail logistics. The GAZ would carry a Soviet-built tactical radio with special frequencies, allowing it to relay coded messages through American aircraft orbiting the Trail. Each of the men was outfitted with an authentic uniform, weapons, and equipment.

My job was to provide them with convincing infiltration packages and critical "pocket litter" that would stand up to rigorous inspection by either NVA security forces or *Trinh Sat* officers they might encounter. Working with our interpreter, I consulted the defector on the exact language required to complete the People's Army of Vietnam identity documents, travel orders, letters of introduction, and ration and fuel coupons. Although the paper had been carefully duplicated in a secret Saigon laboratory, it bore the appropriate scent of jungle mildew that permeated every document ever recovered from the Trail. We had even gone so far as to "season" with rust the stickpins holding bundles of documents together. The rust left behind small telltale orange stains on the sheaves, which looked as if they had been repeatedly opened and stamped by regional security inspectors—exactly the impression we wanted to convey. It had taken me many months to collect the original exemplars and reproduce all the validating stamps and forms needed for this mission.

On this assignment, I was working closely with Jacob, my mentor from the hectic BARGER operation in Hong Kong. As I told Jacob the last time I'd seen him in Saigon, "We've got enough paper for this guy to drive all the way to Hanoi and back."

We were drinking Pernod with a man I'll call Fred Graves, a "bang and burn" special devices officer and chief of the local TSD contingent, on the roof terrace of the Majestic Hotel, overlooking the bend of the Saigon River. Fred and Jacob preferred the French *colon* decadence of the hotel to the rowdy, frat-house atmosphere of happy hour at the CIA's downtown BOQ, the Duc Hotel, five blocks away. Across the river to the south, distant flare planes were illuminating the dusk around a firebase on the northern edge of the Mekong Delta. The weak sounds of distant artillery fire reminded us that we were indeed living in a country at war.

"Currency," Jacob had said curtly. "Never forget you're in Asia, and the type of money this chap will be carrying will make up a vital part of his pocket litter."

"Bet your sweet ass," Fred added. "Charlie likes his money."

We planned to "insert" the defector back in his old role as a lieutenant in the General Political Directorate, and we needed to be certain that both he and his companions were carrying the type of money that would pass muster, be it North Vietnamese dong, Pathet Lao kip, or U.S. dollars. That night, I'd gone to the Station and sent another in a string of urgent cables around the region, requesting "Essential Elements of Information" (EEI) to be collected from Communist POWs and *Chieu Hoi* ralliers undergoing debriefing.

As the operation matured, two document analysts from Headquarters joined me to make certain our infiltration package was absolutely current. After all, the rallier and his Pathet Lao colleagues had been out of the enemy loop for almost six months. In a flurry of trips between the Saigon Station and the secret MACV/SOG base at Camp Long Thanh, my team confirmed that U.S. dollars in twenties and fifties, combined with a token amount of Hanoi dong, would be the most appropriate currency for a political officer to carry. We also verified the latest subtleties of route-pass chops issued up and down the Trail. Armed with this information, we had joined the infiltrators in Laos.

It took me most of the day to complete the paperwork as the three defectors watched warily. Understandably, they were an attentive audience—their lives depended on my skill.

Covert action paramilitary officers then took over, showing the team their GAZ 69 A jeep with its mounted radio. They were already familiar with this vehicle, but the next piece of gear they were shown could have been a UFO for all they knew. In fact, it was the MKWURLY, a powerful, twin-engine variant of the workhorse Huey that TSD had developed specifically for this type of insertion. Normally, Hueys made a characteristic "whomp-whomp" noise due to their wide, two-bladed rotors, but these choppers had been completely silenced by installing a smaller, multi-blade rotor, geared to spin much faster. The pilots had been trained to fly in absolute darkness, wearing infrared night-vision goggles.

As the infiltration team practiced with the American crew, loading and unloading the jeep aboard the strange chopper, we all understood that the operation was, at best, a long shot. But we also knew that Special Ops teams all over Indochina were conducting similar efforts to locate POW camps, where rescues might be attempted in face of the Communist intransigence.

On the night of the insertion operation, I watched the bizarre WURLY lift off silently and head toward the dark mass of the Truong Song. It was now up to the defector team to follow their orders and training so I settled down with the Special Ops officers, waiting for the first radio SITREP (situation report).

Two days later, we received our initial indication of trouble, a message broadcast in voice code from the team leader. One of the Pathet Lao, he said, had decided to redefect to his old unit. A gun fight had ensued between the two Lao members; one was dead, the other wounded. The NVA lieutenant we had worked with for so many months then announced he was turning himself in to the "proper authorities" of the People's Army.

That's probably what he intended to do all along, I thought with some bitterness.

There was, of course, a bright side to all of this. When Vietnamese military intelligence carefully analyzed all the infiltration documents and gear, and heard tales of silent helicopters flying through the darkness to deliver NVA vehicles deep into the fortified heart of the Ho Chi Minh Trail, they must have wondered, Who among us can we trust?

Fred Graves, who had come up from Saigon for the operation, folded his map and dusted his hands in ironic finality. *"Mai pen rai,"* he said, a Thai expression appropriate to the situation. Literally translated, it meant "never mind," but in the context of our often frustrating profession, Graves's words conveyed something closer to the Thai fatalism: *"Mai pen rai krap,"* or "What can you expect?"

Vientiane, Laos, November 1972 ■ Jacob and I were working as quickly as possible while our two subjects conversed intently in rapid, animated

French. The room was stuffy with the blinds and drapes tightly closed, and the table lamps glowing brightly. Lou, the resident TOPS officer, had also attended this landmark agent/case officer meeting with us to record it on videotape. That videotape is still highly classified—I cannot disclose all the details of the disguise techniques developed for this operation, which I'll give the code name GAMBIT. These techniques have evolved over the decades, and many are still employed today.

In the hot glare of the closed room, I worked with one of the subjects, a handsome, young African-American case officer. Meanwhile, Jacob zippered a skin-tight glove onto one arm of the agent (cryptonym PASSAGE), a small Lao who was jabbing the air with his free hand while speaking in a language more evocative of the Seine than the Mekong flowing outside the safe house windows. As he answered, the case officer kept his head motionless while I sewed a light brown, fashionably long hairpiece into his closely trimmed Afro.

The two men paid little attention to us. Their meeting had to be accomplished quickly, and they were completely consumed by the urgency of the subject matter. The local official was passing on the results of the neutralist prime minister's cabinet meeting that day, which included details of Hanoi's secret bargaining position in Paris. The case officer had to respond to an urgent EEI list, forwarded from Kissinger's shop in Washington via Headquarters, by sending an IMMEDIATE cable before morning in Paris.

Jacob and I worked with focused speed on the disguises while trying not to intrude on the conversation. This mission was my first operational trip after undergoing disguise training at Headquarters that summer, and I wanted it to come off without a hitch.

By 1970, I had become chief of a field unit of sixteen graphics specialists. I had proposed the concept of cross-training employees in TSD's Authentication and Graphics branches so that they could be clustered in smaller units closer to the action than our Okinawa base. Senior management approved of my plan, and Jacob and I now represented a prototype Authentication Generalists Program, operating out of a small regional base in Southeast Asia nearer to where we were normally needed. Jacob was the Authentication half of the team, while I fulfilled the Graphics function.

In effect, we had broken the "rice bowls" of some well-entrenched bureaucrats back at TSD Headquarters, but in doing so, we had saved money and manpower. Now we had to prove that we could indeed deliver the autonomous "quick reaction capability" (QRC) we had promised.

We and other espionage services used disguise heavily in Vientiane. This hick town up on the Mekong teemed with spies from two superpowers and a handful of local factions. The case officer I had to disguise was one of the few African Americans in Vientiane, and certainly easy to keep under surveillance. But now he was in desperate straits, having successfully recruited a major national figure who had agreed to provide vital information on the Communist side of the Indochina peace talks.

For several weeks, they had been able to meet in relative privacy at a safe house, using fairly secure car pickups and dropoffs late at night. But with the Pathet Lao drawing closer to town, the local militia declared a curfew and threw up random roadblocks at night. Our case officer knew it would be disastrous if his well-known agent was ever discovered with him. We doubted that the militia would stop a car bearing *Corps Diplomatique* license plates, but PASSAGE was too nervous to take the chance. In fact, his nerves had become so frayed that it looked as if we might lose this vital source.

When I arrived, the situation had become very serious, and I had to work on several pending disguise cases. After Lou took me to meet this case officer, I'd sent an IMMEDIATE cable back to Headquarters asking for advice before attending the nightly meeting with PASSAGE. Head-quarters responded at once, requesting I make copious measurements of the two men's heads and photograph them from every angle.

I then prepared the details of my plan, which involved the Agency's new consultant in Hollywood, whom I'll call Jerome Calloway. Calloway had recently started working with the Headquarters Disguise Unit, and he was then at the apex of his career, having just received top industry awards for his makeup work in sci-fi movies. When I had been at Head-quarters that summer, my Disguise colleagues were perplexed. How could they use Calloway's products in the real world of espionage? Now, I suggested a concrete plan for our Vientiane case officer and his agent, PASSAGE: Why not use the more malleable disguise materials for their faces, hands, and lower arms? These had completely transformed the Hollywood actors in Calloway's films and could alter the appearances of our own case officer and PASSAGE in similar fashion.

By pure coincidence, the measurements of the case officer and PAS-SAGE closely matched the dimensions of materials Calloway had al-ready made for the stunt doubles for Victor Mature and Rex Harrison. Headquarters promptly sent us these materials.

PASSAGE and the case officer had arrived undisguised for this vital debriefing just after dark, but we felt it would be crucial for them to leave fully disguised.

Halfway through our job, we stopped and stared at the subjects. Then they fell silent and gazed at each other. Jacob and I had success-fully transformed an Asian statesman and an African-American case officer into two Caucasian diplomats vaguely resembling the two Holly-wood celebrities.

"Good evening, sir . . ." PASSAGE said in heavily accented English to his case officer. His little joke broke the tension in the room.

As they left the safe house that night, the embassy car rounded the corner and encountered a militia roadblock that had not been there five hours earlier, but the nervous young troops waved the CD car through the barrier without hesitation. Our disguises had passed the first test.

Over the coming weeks, we taught the team to use the materials quickly and efficiently in the car as it moved through the darkness. Sometimes, PASSAGE slipped into the OPS vehicle as it coasted through the shadows near the gaudy arches of Patuxai, Vientiane's equivalent of the Arc de Triomphe. Between the time the car left the traffic circle and approached the first roadblock, both the case officer and PASSAGE were in full disguise.

Obviously, these somewhat crude features would never hold up under close inspection, so Lou took photographs of the two men in disguise and we produced alias Lao diplomatic *cartes d'identités*. Whenever they came to a roadblock, the men simply flashed their ID cards, and the ragtag soldiers saluted the personages aboard the gleaming Buick. Listening to their accounts later, I recalled the high school dance that Doug and I had crashed as kids in Denver. I had learned then what I knew all too well now. Successful deception involving disguise was as much a matter of planning, demeanor, and attitude as of visual appearance.

Again and again, that resourceful American case officer and courageous Lao gave Kissinger's delegation in Paris vital insight into Hanoi's bargaining position in the Indochina peace talks.

Several months later, Jacob and I were thrilled to hear from Headquarters that during budget hearings on the Hill, Deputy Director of Central Intelligence William Colby had used our breakthrough case as an example of the Agency's innovative spirit. Fortunately for us, this mission was just the beginning of Calloway's effort to help us rewrite the book on the CIA's disguise tradecraft.

Bangkok, Thailand, March 1973 ■ It was from Bangkok that I watched the release of American POWs from Hanoi on Armed Forces Television with a mixture of pride and deep remorse. Fewer than seven hundred came home, out of more than a thousand we felt certain the enemy had captured alive in Vietnam or Laos.

Ultimately, America's costly involvement in Vietnam was a tragic defeat. From the perspective of an intelligence war, we had failed to understand the fundamental nature of the enemy. Successive administrations and CIA leadership could only perceive the North Vietnamese through the lens of the Cold War, as surrogates of their Communist masters in Moscow and Beijing.

That assessment had been partially true. As a young man, for example, Ho Chi Minh had been one of the founders of the French Communist party. But the North Vietnamese were also fervent nationalists, absolutely determined to unite all of Vietnam under their control. Few leaders in Saigon felt such fervor about protecting their own country.

In hindsight, it is obvious that the Vietnamese Communists could not overrun all of Indochina, knocking over "dominoes" all the way down the Malay peninsula and throughout the Indonesia archipelago. But in 1965, when Lyndon Johnson made major military commitments to the defense of South Vietnam, the situation was far murkier. As the war unfolded, a major responsibility of America's civilian and military intelligence was to provide accurate evaluations of enemy strength and intentions to Washington.

The system failed. Our policies had been shaped by preconceptions. That was a luxury in which political leaders often indulged, but to which intelligence professionals should never succumb.

For me, however, those experiences in Indochina laid the groundwork for other major Cold War engagements that provided the pivotal successes we needed to help tip the balance in our favor.

5 ■ Kipling's Beat

It is not good for the Christian health

To hustle the Asian brown

The Asian smiles, the Christian riles

It weareth the Christian down

The end of the fight is a tombstone white

With the name of the late deceased

And the epitaph drear "A fool lies here

That tried to hurry the East."

—Rudyard Kipling

Southwest Asian Seaport, Early 1970s ■ The first time I saw the subcontinent, I was on an urgent exfiltration case. Flying into the international airport, "David" (my fellow tech ops officer) and I had to be especially alert because we were traveling as tourists and not carrying official passports. In this region, if we were caught assisting the escape of a Soviet defector, we could go to jail.

The Air France 707 shuddered as we descended through the turbulent darkness toward the runway. I gazed out the window, expecting to see the lights of satellite towns and highways near the large seaport. After all, this was one of the most densely populated regions of the subcontinent. But besides the occasional crawling vehicle light, the landscape was pitch black.

The plane touched down and slowed. It seemed as if we were completely engulfed by the "smit," which Jacob had warned us would be especially repulsive this season. Smit was an acrid pollutant when mixed with diesel and two-stroke jitney scooter exhaust in a sprawling city like this.

Indeed, my eyes were watering as David and I edged our way through the humidity and crowds enduring that local form of torment known as Arrival Formalities. I had grown accustomed to East Asia and Indochina, where people kept a physical distance from Europeans, if not from each other. But in this even more congested part of the world, the concept of personal space did not exist. The people here certainly thought nothing of assaulting each other with their limbs and luggage.

As we made our way across the terminal, I tried to ignore the heat, the wafted stench from the W.C.s and open sewers, and the voracious flies. As in any foreign airport, it was our job to analyze the local security controls. But we had reason to be especially vigilant here: If all went well, the operation would bring us back to this airport with the defector in a few days.

Experience had taught me that a practiced eye could ascertain the pecking order among the bureaucrats at any international border. Here, Immigration was clearly at the bottom of the heap, with Customs at the top. Security was somewhere in between, overtly represented by stern policemen in starched khaki shirts and shorts, with military puttees over their bare calves and sandaled feet. They brandished wooden batons and projected that aura of tired scorn common to police from Moscow to Sydney.

But Jacob had informed us that the uniformed ranks of the Ministry of Interior were of little concern to us. Our "real opposition," he had cautioned, would be the plainclothes officers from Special Branch (SB), responsible for national security and counterintelligence. Several had to be in the terminal now, watching the passengers arriving from international flights, but I did not know how to spot them in the exotic mixture of saffron turbans, sweaty gray Nehru jackets, and travel-worn purple saris. This area of the subcontinent was obviously a complex human matrix, and I was filled with curiosity.

I had to remind myself that David and I were not here for cultural enrichment—we had pressing practical matters to attend to. We would be briefed at the local CIA base on the subtleties of the Special Branch before the actual exfiltration, so I wasn't too worried.

We had valid tourist visas and diligently completed all the entry forms presented to us. Immigration did not probe for what our real purpose might be beyond the word "tourism" we both scrawled on the paperwork.

Our purgatory in the Immigration line was relatively brief compared to Customs, whose inspectors knew their jobs very well. They didn't seem particularly concerned with Westerners, but they were obviously well-trained in spotting their own citizens who might be of particular interest. The locals were among Asia's most notorious smugglers; here at the country's commercial heart, the incoming and outgoing illicit flow of gold or hard currency was the main attraction. Consequently, customs inspectors meticulously examined travelers' airline tickets and studied the immigration cachets in their passports to decide whom to question in detail, how many bags to open, and whether to subject the unfortunate soul to a secondary inspection in the airless back room.

With the local currency officially pegged at eleven to the U.S. dollar, but going for twenty to one in Hong Kong and Bangkok, smuggling their own money was quite lucrative. Large denominations of the more desirable and compact U.S. dollars were an especially handy way of conducting business in this country, where profits could be exported to overseas accounts beyond the reach of tax collectors.

While I amused myself trying to memorize the exact sequence of formalities in this sweltering mob, David was clearly on the brink of losing his self-control. He was one of the best documents men I ever worked with, but he liked his creature comforts, and there was little comfort that night in the stifling arrivals hall. Neither of us had slept much in the last day and a half. The jostling, heat, and stench were annoying to both of us, but we couldn't leave the line; the currency form was mandatory. Although it seemed to have been printed in microscopic type on smeared newsprint, it had to be completed in multiple copies, using worn-out scraps of carbon paper. Every time we changed dollars during our stay, we had to have this form stamped by another platoon of bureaucrats. The most important aspect of this whole frustrating procedure from our point of view, however, was that Customs did, in fact, require this crucial document form to be processed both upon entry and departure from the country.

More than two hours after we arrived, David and I had completed all the formalities, only to discover we had yet another painful process to endure, Tourist Assistance. We hadn't had time to book hotels in advance (and certainly didn't want the local CIA base making reservations), so we had to fall into yet another line.

It was almost four A.M. when we got our hotel reservations, retrieved our bags, and left the terminal. The turbaned Sikh who ran the official taxi rank stepped up and directed us to the cab at the head of the line.

I noted that the car that slid in behind it, a yellow-and-black Hindustani Ambassador, an almost comic caricature of a 1950s sedan, was driven by another Sikh. I graciously allowed three Japanese tourists behind us to take the first car, while David and I piled into the second. Call it paranoia or good tradecraft; traveling as we were, it was not wise to take the first taxi assigned to us.

If our driver was an SB agent, he lived his cover very well. Tall and lean with a splendidly waxed handlebar mustache and a full black beard tightly rolled into a chin net, he deposited our bags in the boot and made sure we were safely seated before climbing behind the steering wheel.

"Panorama Hotel!" I shouted in response to the driver's mumbled question. He drove with wild élan, leaning on his chirping little horn and bellowing curses at other vehicles swarming in the road around us. While David tried to sleep, I watched the bizarre spectacle emerging slowly from the smit as the road raced by. Occasionally, we overtook a lumbering truck laden with gunnysacks, or an oxcart with tall wooden wheels that shrieked against the rough concrete. Except for the motor vehicles, I knew that the scene around us had not changed for centuries.

Our slow progress gave me time to contemplate the urgent nature of this exfiltration mission. It had started like so many other similar assignments, with a late-night phone call announcing that an IMMEDIATE NIACT cable had just arrived and been decoded. "Shower, shave, and pack your bag," my Okinawa boss had said. "You and Dave are going on a trip."

En route to the Naha airport that morning, I reflected on the IM-MEDIATE that Jacob had sent. This was not just some routine exfiltration of a Polish embassy chauffeur or a boozy Czech poultry expert. Instead, a young but uniquely placed KGB First Chief Directorate (Foreign Espionage) officer, stationed at the big Soviet embassy in the diplomatic capital in the north, NESTOR, had decided to defect to the United States twelve days earlier. Working under innocent attaché cover, the man had been given relatively free access to the capital, unlike many of his colleagues who could only leave their embassy compound guarded by security officers.

NESTOR had told his associates he was going to watch a film in an air-conditioned cinema with local friends. In fact, he had been planning this fateful step for months. Making the decision to completely turn your back on your previous life, on everything and everybody you have ever known, was as comforting as stepping over a cliff and falling into oblivion. Once you had taken that first step, it was almost impossible to return, and I understood the danger of the situation into which NESTOR had willingly plunged.

Part of NESTOR's duties as a KGB officer required that he know the identities of officers at the local CIA station. It was to one of their homes that he went that night, not to the air-conditioned cinema. The American case officer went by the book, requesting that NESTOR return to his embassy and obtain something, a document that would establish his bona fides. Once they were confirmed, the Station would give him instructions for primary and secondary contact sites elsewhere in the country where the exfiltration would begin.

NESTOR had arrived at the case officer's house with only a national diplomatic identity card describing him as an attaché, hardly material for gearing up a crash exfiltration operation involving the CIA station, bases, and a team of TSD experts. But NESTOR was back at the case officer's house the next night, bearing intriguing samples of Soviet cable traffic and dispatches. The American officer was convinced beyond any doubt that he had a potentially invaluable defector on his hands. Such a "walk-in" was every case officer's dream. The

American was ready, presenting NESTOR with clear instructions for primary and secondary contact sites, with multiple fallback dates and times for NESTOR to meet with TSD specialists and prepare for the actual exfiltration.

The American also realized the KGB and the Special Branch would scour the city for NESTOR when he did not appear at his embassy in the morning. The KGB was thin on the ground compared to the SB, but the national security forces could virtually saturate every public transport facility in the country in a matter of hours. The American officer offered to move NESTOR to a temporary safe house until secure travel to a contact site could be arranged.

NESTOR declined the offer. "I've given this a lot of thought," he said in his fluent, American-accented English. "I'll rendezvous with your people in ten days." He carefully wrote down the date, time, and place in the southern port for the meeting.

Being a competent, Moscow-trained case officer in his own right, he went to ground, drawing on his own resources, a prospect he seemed to relish. Clever and confident, NESTOR could disappear into the local masses and make his way by land to the port, using a convincing native legend. Already tanned from months on his embassy's volleyball court, and naturally small in stature, he shaved his head to complete his metamorphosis into a native. He bought a cheap bag and loose cotton clothes in the bazaar, then headed off northeast by train, in the opposite direction of his final destination. Having run several nets of local agents, he had two virgin sets of national identity papers.

The first set took him by crowded second-class railway carriage as far as the pilgrimage sites. From there, he doubled back by village buses, finally taking a night train to an industrial town, where he spent three days holed up in a cheap rooming house, living off tinned food and tea cooked over an alcohol stove. On the fourth morning, dressed in baggy white cotton shirt and trousers, he bought a ticket on the late afternoon express to the port, using his second and last set of identity documents.

These precautions had not been too elaborate, given the aggressive nature of the pursuit that the KGB and the Special Branch had mounted. When NESTOR had disappeared from the Soviet embassy, the KGB *resident* had pulled out the stops, flooding the airport and all the train stations with security men. By the end of that day, the SB had covered every bus station in the teeming capital. The next morning, newspapers across the nation ran stories announcing the "disappearance under mysterious circumstances" of a young Soviet attaché. Each story bore an excellent photograph of NESTOR.

Because of the security situation, all passengers leaving the country by air were instructed to reconfirm their flight in person twenty-four hours before their departure at a local travel office. If this order was disregarded, the traveler would not appear on the flight manifest and could be detained at the airport. Essentially, NESTOR's defection had triggered one of the most intense and comprehensive security operations in the modern history of this region.

NESTOR was undaunted. When he arrived at the contact site in the port city on the appointed day and hour, he displayed the proper recognition signal—a rolled up copy of *The Times* in his left hand—and used the parole, or code word, "Igloo." The case officer was taken aback by NESTOR's appearance. The Soviet had become just one of thousands of other mixed-blood or upper-class locals, dressed in a flowing white Nehru shirt and a black lamb's wool cap. It was easy to see how NESTOR had evaded the combined KGB and SB search for the previous ten days.

The CIA base officers immediately whisked NESTOR to a safe house south of the city. They then sent cables requesting Jacob to assemble a team and arrive "soonest" to help mount a risky but vital exfiltration operation.

During NESTOR's initial debriefing at this safe house, it became clear what a potentially important defector he was. Although only in his thirties, he had spent most of his life in the KGB. Coming from a well-established "Chekist" family with roots far back in the Soviet security services, NESTOR was the first defector from the group that the CIA's Soviet–East European Division had dubbed the "Junior KGB." Under aliases, NESTOR had spent some of his teenage years as the "son" of a Soviet embassy attaché in London, and then several more years as the nephew of a GRU (military intelligence) officer in Washington, attending Bethesda–Chevy Chase High School. He therefore spoke both British and American English without an accent. Back in Moscow, he attended KGB institutes, emerging with advanced degrees in Asian languages, including Hindi and Urdu. He was on his second tour at the *residentura* when he defected.

For the CIA, NESTOR represented a treasure trove of information. Not only could he identify the entire generation of "juniors" being trained for overseas assignments, but he could also give us a virtual "wiring diagram" of the KGB's vast Afghan, Sri Lankan, Indian, Pakistani, and Burmese espionage operations. Finally, before leaving the embassy, he had collected a nice sampling of classified messages and code-pad materials that would keep the computer wizards at the National Security Agency busy for years.

Safely exfiltrating NESTOR from the tightening KGB and SB noose was a challenge worth the risk.

Unable to doze off in the back of the jolting taxi, I was once again pondering all this information as we entered the outskirts of the target city. We passed endless rows of people sleeping on rope mats scattered on the sidewalks, their cotton garments pulled up over their faces like cadavers. Others were too poor even to afford the thin mats and slept directly on the garbage-strewn streets as goats foraged around them. These chaotic streets gave way to hivelike clusters of shacks next to towering, mildewed tenements and office buildings. I had read that half the population of this city lived in the streets, but I hadn't been prepared for the visceral reality. Each time we stopped at a traffic light, the taxi was surrounded by beggars, young and old; some were lepers with deformed faces and hands, pawing at the windows. The driver shouted at them angrily, waving a fly whisk. In my fatigue, the wretched creatures appeared like phantoms from a Kipling novel and, while I was intrigued by what I saw, I was relieved when we finally reached our hotel a little after dawn.

Maintaining our cover, David and I spent ten minutes selecting tourist brochures from the rack at the reception desk before stumbling like zombies to the elevator, looking forward to a well-deserved shower and breakfast before making our first contact call to the local CIA contingent. To the SB observers in the hotel lobby, we were just two more jet-lagged Westerners intent on absorbing as much local "culture" (including visits to high-class bordellos and casinos) as was humanly possible during a short stay.

CLEAN AND FED, but not yet rested, David and I waited in our room until exactly two minutes after nine. Then I called the local contact number, which we had both memorized from the cable.

"Is Sally in this morning?" I asked, as also specified in the cable.

"Not yet, I'm afraid," a woman answered in a neutral accent that could have been Canadian or American.

As we left through the lobby, David and I stopped at reception again. "If the Nepalese tourist bureau calls," I said earnestly, "please be sure to take their message. We're hoping to do some trekking in the Himalayas."

The efficient Eurasian jotted down some precise notes on his pad. "Without fail, sir," he replied with Victorian formality. My intuition told me that this tidbit of information would be in the hands of Special Branch before we were a block away.

Clutching tourist maps of the city center, David and I crossed several busy streets, carefully avoiding heaps of garbage and piles of construction debris sprouting shoulder-high weeds. We edged our way through the colorful crowds, which displayed every imaginable stature and complexion, dodging Bedford trucks and orange-and-white Russian buses, all belching diesel smoke. Performing a surveillance detection run in these conditions was simple because we could rely on the ploy of losing our way in the unfamiliar streets and then doubling back.

At 9:45, we had reached the rendezvous, an inconspicuous spot in front of an electrical supply shop on a bustling street. A tan Datsun sedan, driven by a young, red-headed American in a wash-and-wear suit and narrow blue tie, pulled up to the curb and paused only long enough for us to jump in. As we turned off the street, I noticed the elegance of the pickup spot, the ideal place for a rendezvous. It was on a blind curve that could not be directly observed from either corner. Only blanket stationary surveillance posts along the entire street could have spotted us.

The driver took us for a confusing ride through a labyrinth of narrow streets, bursting with honking trucks and three-wheeled jitneys sagging under people and cargo. We crossed a wide boulevard and entered another network of lanes, where dhotis, Muslim beards and turbans, and pajamalike shalwar kamez glided by.

As we made our way toward the hot blue expanse of the sea, I grasped the true purpose of all this elaborate tradecraft. For more than ten days, the KGB/SB hornet's nest had been stirred up by NESTOR's disappearance. By now even the lowest-ranking SB officer had to realize a major security problem was unfolding. For senior officials in the Special Branch, NESTOR's possible defection was a calamity: NESTOR knew where the "local bones" were buried. If the CIA managed to exfiltrate him, we would have gained extremely compromising information. Without question, the SB was as motivated as the KGB to stop us.

The red-haired driver introduced himself simply as "Mac." From his knowledge of the local geography, I assumed he had been assigned to this area for at least a couple of years. After a fascinating tour of the old quarter, we turned onto a seafront boulevard and drove past a long line of mildewed, whitewashed apartment buildings.

"This part of the beach is famous," Mac explained. "On weekends, and during big religious festivals, you might find three or four million people along this road."

He added that these festivals were an ideal time to have brief encounters with assets because the staggering multitude of people shouldering their way among the snake charmers, pony rides, magicians, and food stands selling spicy dishes made surveillance nearly impossible.

Mac then briefed us on the status of the operation. NESTOR was still hidden at the safe house with Jacob and a preliminary debriefer from the Soviet–East European Division, who had flown in two days earlier. The vigilance of KGB and SB stakeouts at the international airports and border crossings had intensified in anticipation of NESTOR's attempt to escape. SB surveillance around Western embassies and consulates had also increased. Our tentative launch date and time for NESTOR's exfiltration was April 22 on TWA's westbound around-the-world flight to Athens, scheduled to arrive at one A.M. and depart an hour and a half later.

That gave us three days to solidify and refine the final exfiltration ops plan, as well as create the false documents and disguise package. We would stay clear of the CIA base, working instead in a commercial cover office, where Mac and his boss, Raymond, would finalize the local end of the ops plan and handle communications. Jacob would put the finishing touches on NESTOR's disguise at the safe house, while Mac would be the key cut-out, or liaison, between the office where David and I would work and the safe house where NESTOR was hidden. Therefore, Mac had to be particularly careful on his trips to and from the safe house, changing cars in discreet garages, using circuitous routes, and altering his profile with an array of hats and hairpieces. A simple one-way trip from the office to the safe house might take as long as three hours—an inevitable nuisance, but the price of good security in any case.

Since David and I would be involved in the airport departure phase of the exfiltration, we had to maintain our tourist cover, stay "clean," and could not be seen again with Mac, because he might soon come under surveillance. Further, Mac explained, we had to immediately change hotels because the Panorama was frequented by East Germans and Czechs—and almost by definition, the KGB. These gentlemen would be on a high state of alert, with their antennae searching for any suspicious Americans who had just coincidentally arrived in town while NESTOR remained free. In their present state of anxiety, the KGB was hardly likely to dismiss coincidences, and we had to remind ourselves of the power of Murphy's Law.

Only three days, I thought, trying to shake off my fatigue and plan my next steps in the operation as logically and efficiently as possible.

Mac left the seafront and turned onto a winding road that snaked upward through a densely forested hill. A stone wall on the left enclosed a lush park, but there was a foul stench in the air. Shadows crossed the windshield, and I looked up to see hundreds of vultures circling above the trees.

"This is the Parsee cemetery," Mac said, and went on the explain that the Parsees were descendants of Persian Zoroastrians, who believed earth, water, and fire were sacred and should not be defiled with human remains. They therefore placed their dead on stone pillars in this preserve, where the vultures came to feed each day.

"Cheerful place," David muttered, rolling up his window to shut out the repulsive breeze from beyond the wall.

This detour along this desolate road assured Mac we were not being followed. He dropped us off near the base of the hill so that we could emerge onto a busy street brandishing our tourist maps and catch a taxi to the neighborhood of the cover office. We asked the taxi driver to stop at a travel agent office, two blocks short of the address Mac had given us, where we picked up some dusty brochures about the old Himalayan hill stations. Then we strolled up the street and turned abruptly into the office building entrance, right past the *chokadar* gatekeeper, as if we entered this lobby every day.

The base maintained a commercial cover office in this busy building as a convenient site for "nonofficial" contacts. The port was a regional business center, and it wasn't unusual for foreign bankers and commercial reps to visit throughout the year. The fact that this particular block of offices stood among thirty similar buildings intensified its anonymous nature. The nearby sidewalks and lobbies were well salted with British expats, Europeans, and a fair number of Americans. Even the vigilant Special Branch and the KGB could not monitor every foreigner passing through this district.

Following Mac's instructions, we rode a creaky old elevator to the top floor of the six-story building, three floors above our actual destination. After taking the stairs to the third-floor landing, we waited several minutes to make sure no one was watching us before we entered the cover office at the rear of the building, with a convenient escape route across the flat concrete roof of an adjoining office block. Mac introduced us to "Raymond" and "Jane," the local case officers who had been working virtually nonstop since NESTOR arrived. Raymond, the senior, asked Jane to get on the phone immediately and find us new hotel rooms near the central city beachfront. Twenty minutes later, she came back to the rear office shaking her head.

"No good," she said. Apparently there was an international agricultural equipment exposition in town, and wealthy farmers had come down to enjoy the beach as the heat mounted in the interior. Then, Jane remembered an old beach cottage that expat friends kept out near the Southern Paradise Resort, west of the city. She made one phone call and secured the cottage for us. We felt confident with her choice; it was unlikely that the SB would spread their net that far.

Despite the unappealing prospect of vacating a comfortable, air-conditioned hotel room in the afternoon heat, David and I sat down at a table with the case officers, braced ourselves with several cups of soothing tea, and began to flesh out the exfiltration operations plan. The initial planning lasted all day.

The next morning after a fitful, mosquito-plagued night at the beach cottage, we employed more surveillance evasion procedures on the way into town, around the office block, and up the elevator. Once safely inside with Mac and Raymond, we tackled the serious work of preparing NESTOR's forged travel documents.

Since the case was so challenging, and failure unacceptable, we had to prepare more than one option for NESTOR while we awaited Headquarters' final approval on our ops plan. At some point during the debate that was now raging between Headquarters and the field, everyone would agree on a primary and secondary alias identity for the subject, as well as a primary and secondary exfiltration route. We could then begin entering the "back travel" cachets in the passports NESTOR might use and "issue" some of the backup documents. But we couldn't stamp the forged cachets into the passports indicating NESTOR's ostensible route into the country, nor could we complete the bio-data pages, until we knew what he would look like after Jacob transformed NESTOR into his disguised alias persona.

Raymond spread a sheaf of cable traffic on the table, which indicated that the discussion between all concerned continued unabated. No one could decide exactly how this exfiltration operation should proceed, and later, after I had run a dozen of these harrowing jobs, I would come to call this state of paralysis the "committee effect." The cause of this phenomenon was a matter of both tradecraft and trust: No experienced case officer could accept the idea of turning an untested asset, especially a very important agent or defector, loose in an airport under any circumstances. It didn't matter that NESTOR himself had been trained by some of the world's premier street espionage experts at the KGB's Red Banner Institute for the First Directorate, or School 101 for spies, in Moscow.

There were factions in every operation—and this case was no exception—who believed that slipping the subject into the trunk of an official American diplomatic car and driving him out of the country "black" (a scenario in which the American spy, not the asset, would do the talking to get them out of a jam if challenged) was the most effective course of action.

"Sounds good in theory," Raymond grumbled. "But there you stand, with the American flag draped all over you when the border police open the trunk. What happens to plausible deniability if the guy's a KGB dangle and sings to high heaven?"

Jane poured more tea to calm us down, but she, too, seemed anxious. "If the SB even suspects NESTOR is hiding on American diplomatic property, they'd march right in and grab him, the Ambassador be damned, so they sure as hell wouldn't think twice about opening the trunk of a sedan with U.S. dip plates."

Their point was well taken. The KGB would turn this into a major international diplomatic and media scandal, then proceed to send even more bogus defectors against us to score more points. The case officers on the scene argued persuasively against the land route using an official car, and we then waited on the final word from Headquarters that morning.

But we were unable to ignore other black options. "I suppose the special ops types will try to sell us on a submarine or a helicopter," Mac quipped.

I saw his point, having spent the past few years among the Agency's paramilitary operators in Indochina. They were doing a fine job there, but in this part of the world it would be enormously amusing to the Soviet opposition and the world media if the local navy discovered a U.S. submarine run aground on the nearby shoals or an "oil company" helicopter down on the beach with engine trouble. It was therefore clear to me that the black option simply wasn't appropriate for NESTOR's case. But later that year, this type of solution did prove to be our salvation when a Communist Chinese diplomat, stationed in a remote South Asian country, defected to us. There were few roads in and out of the more mountainous region, so we borrowed the local USAID helicopter to fly our man to an isolated tiger-hunting camp. We provided him with a disguise and an alias there, then moved him aboard an elephant down through the rain forest and over the border, where he was met with a Land Rover and was eventually sent on to the West by quasilegal means.

Whether the route is commercial or black, an essential goal of any exfiltration operation is to break the trail so that the opposition has no idea what happened. It is always more desirable when the escapee simply vanishes forever; besides disquieting the opposition, such a clean operation also plants the idea among the enemy that they, too, can avail themselves of our services if they wish to defect. In the case of NESTOR, if he ended up missing without a trace, there would definitely be a few heads rolling in the local KGB *residentura*, and retribution might extend to the SB as well.

A quasi-legal departure, effectively using TSD wherewithal and a good operations plan, on board a scheduled airliner (preferably a U.S. carrier) bound for friendly territory, presented the style, efficiency, and finality we sought. Over the coming years, as I went deeper into sophisticated exfiltrations, our motto was "There's nothing so exhilarating as hearing the wheels come up." We found that we could always grab the attention of local case officers by presenting this argument: "Exfiltrations are like abortions—you don't need one unless something has gone wrong. But if you do need one, don't try to do it yourself. We can give you a nice clean job." It was precisely the kind of tongue-in-cheek bravado that case officers loved, and it served as our distinctively audacious calling card. But there was a serious underlying principle involved. Ideally, an agent would remain productive, and a walk-in defector such as NESTOR would become an agent-in-place; neither would require exfiltration, unless their cover was about to be blown or they were under hostile pursuit. At that point, "coming in from the cold" meant something had gone very wrong.

Since the NESTOR operation was my first "exfil," I was about to learn from more experienced colleagues, principally Jacob, that running such an operation properly is one of the major criteria by which the professional level of an intelligence service can be judged.

Historically, the Nazi Abwehr and the prewar Soviet NKVD had been past masters at infiltrating and exfiltrating illegal agents. The various Jewish resistance underground groups that arose during the calamity of the Holocaust and its postwar aftermath in the Israeli war of independence had learned fast, hard lessons about illegal border crossing; by the 1970s, the Mossad was the best in the business at moving people in and out of denied areas. The raid on Entebbe will always serve as the premier example of a perfectly planned and executed rescue operation, which is basically what an exfiltration boils down to.

Soviet intelligence had always worked steadily to perfect the practical methods and demanding techniques required in the illegal movement of officers, agents, and defectors. The CIA, via the OSS, had come late to the business. It was a matter of honor that we learn from our mistakes and overtake the field. After my initial lessons in South Asia, I became part of our effort to maintain a perfect record of success, and helped conduct more than 150 such operations during twenty-five years of service in positions ranging from junior TSD Graphics artist to senior Agency officer. That flawless record remains intact, to the best of my knowledge.

This achievement was not incidental. We were always ready for the day we would have to bring one of our assets (and their extended families, if necessary) in from the cold. Although many assets did not escape because they had already been arrested or compromised before we could try to rescue them, once we undertook to do so, it was an effort that never failed. The inventive "clandestine means" we used to survive urgent situations have often been disparaged in the press, but they were a source of great pride and confidence for us.

Correctly appraising the strengths and weaknesses of border controls, such as those employed at the airport, was vital to overcoming the exfiltration challenge. David and I had just obtained the most current knowledge of the entry procedures. We would have to rely on the experiences of the local officers and Jacob for the exit formalities, acknowledging the possibility that we might have to run a prober through the exit controls at the last minute to determine if there were any new wrinkles. While awaiting word from Headquarters on the cover and route issues, we began preparing two sets of documents for NESTOR, one to be used at the international airport for the primary option, and the second to function as a fallback set, which could be used elsewhere if a final reconnaissance of the terminal revealed an unacceptably high surveillance presence.

NESTOR turned out to be an ideal subject. He was fluent in English and spoke excellent German, allowing us to choose from a variety of nationalities. TSD Headquarters had sent an adequate supply of third-country documents, including passports, driver's licenses, and national ID cards. We also had plenty of "window dressing," such as business and personal correspondence and stationery, and the absolutely essential pocket litter, supplied by the local base, to which we added domestic travel materials, such as the brochures David and I had collected at the hotels and travel agencies. Since the bio-data pages of the passports and identity documents were blank, we had *tabulae rasae* on which we could build a new persona—a rare advantage in defector cases, where time constraints and remote locations usually forced us to play the few cards we were dealt.

Once Headquarters approved an option, however, we'd have to address the problem of airline tickets in two identities, showing the ostensible cover itineraries for travel into the country, as well as the primary and secondary out. But, given local currency restrictions and the intense surveillance of airline offices, we couldn't risk purchasing tickets here. We needed time to arrange a "hand carry" from another city along the route of NESTOR's cover legend. Having such a valid swatch of tickets provided the details to which local customs officials would be extremely alert. Since NESTOR might very likely come face-to-face with a KGB security goon who knew him, creating an effective disguise that would hold up to the relentless heat, as well as the physical and emotional stress of his passage through the airport, was an especially daunting task for Jacob. The disguise had to use a minimum of "spooky" materials, which were in danger of coming undone at the critical moment. At the same time, Jacob had to make significant changes in NESTOR's most distinctive characteristics; a great deal of the challenge lay in Jacob's ability to gain NESTOR's confidence and convince him that he could adopt a completely new personality that would not appear wooden or contrived under stress. This would require Jacob's most astute assessment of NESTOR's true self, which could only be achieved by quickly establishing a bond of trust.

LEAVING TOWN THAT night, our taxi dropped us at the bazaar near a red brick Mogul fort, a combination of a multistall junkyard and a counterfeit luxury-goods assembly line, where we were offered "genuine"

Gucci or Chanel purses for five dollars U.S. Again, this break in the taxi ride gave us a chance to make sure we were not under vehicle surveillance. While David purchased mosquito coils, I bought some Agfa portrait paper so that the photos I would print for NESTOR's documents would appear to have been made in Europe.

When we entered the delicious air-conditioning of the Southern Paradise Resort lobby for dinner, I was struck by an idea. Neither of us wanted to feed the mosquitoes in the stuffy little beach cottage again.

"Let's see if they have a room," I suggested.

David shook his head in dejection. "Already tried."

But I strode confidently to the front desk. The haughty young clerk shook his head before I finished my question. "I am truly sorry, sir. We have nothing available, and unfortunately, there is no chance."

I slid my hand over the polished marble, palm down. "My friend and I are going into the bar for a drink. Perhaps there will be a cancellation." Opening my fingers slightly, I revealed a rolled banknote, the equivalent of ten dollars.

The money vanished with a subtlety that would have impressed a professional magician. "Certainly, sir. Please check with me shortly."

An hour later, David and I were installed in a VIP suite facing the beach, having enjoyed real showers featuring unlimited hot water. I poured the scotch, while David tried to find an English-speaking channel on the antique television set. All we had to watch for the moment was a local quiz show and a scholarly lecture given by a Muslim professor. We decided it was time for dinner.

Having again feasted on the seafood buffet, accompanied by a chilled bottle of Portuguese wine, we both felt invigorated. The nearby nightclub seduced us with the mesmerizing beat of drums and tambourines. Groping our way through the sultry darkness, we were fascinated by the voluptuous young bellydancer on stage, an Anglo-Indian by the looks of her, shimmying in the spotlight. All around us, the eyes of Asian gentlemen dressed in conservative European suits were fixed on the woman. Her stage name was Heather, according to the blaring marquee, and she peeled off one veil after another, finally revealing her generous endowments as she danced in only the skimpiest halter and G-string. The spectacle was certainly racy stuff for the subcontinent, but I imagined that the luxury tourist hotel's location on the outskirts of the city gave the nightclub a more permissive atmosphere.

The small band picked up the tempo and Heather skirted the edge of the stage, trying to lure one of the customers to join her in the spotlight. They steadfastly refused.

Then to our surprise, she arrived at our table. Since we were supposed to be tourists, I figured we should act the part. I was up on the stage gyrating with Heather when she suddenly loosened my tie and began to unbutton my shirt. Leaning close, she called over the noise of the drums, "This doesn't bother you?"

I felt her cool hand sliding over my chest. "Not really," I lied.

Then she leaned even closer and spoke into my ear. "Asian gentlemen can't do this kind of thing, you know. At least not in public." Before the song finished, my jacket was off and my shirt was peeled down to the waist. The crowd was aghast, but no one had left the room.

Back at the table, Heather extended a whispered invitation for me to visit her dressing room. An alarm bell rang in my head—maybe I was being paranoid, but for all I knew she could have been part of an elaborate, hidden-camera scheme. If I went to her dressing room, she might strip to obtain compromising photos before I could bolt. Then she would

J. C. Tognoni at the time of his gold strike in Goldfield, Nevada, 1903.

Mother and father in Eureka, c. 1938.

The Mendez family in Caliente, Nevada, c. 1953: (right to left) stepfather Arch Richey; brother John; mother; sisters Cindy, Maureen, and Nancy (Joey is missing); and me.

Maj. Gen. William J. Donovan,
U.S. Army.

The director of the OSS
in wartime disguise.

The original OSS buildings at 2330 E Street NW, CIA Headquarters from 1947–1964.

Embarking on my first trip into enemy territory in Laos in 1968.

A Hmong "striker" and his family at Lima Site.

Collage of memorabilia showing two aliases and one real name.

Posing as "Robert J. Violante" in Red Square, 1976.

The railroad bridge over the Moscow River, where "Mary Peters" was ambushed by the KGB while making a timed drop to TRINITY.

The cobblestone concealment, shown open, in a niche in the bridge.

The cobblestone concealment, holding subminiature film cassettes, gold and silver, ampoules of medicine, and Russian currency, that Mary placed for TRINITY.

Microfilm instruction used by TRINITY to place the signal for the drop to be made—a lipstick mark on a light pole. (The KGB placed the signal instead.)

Translation of the note in the concealment designed to thwart anyone who accidentally found the dead drop before TRINITY could retrieve it.

The ladder from beneath the railroad bridge used by the KGB to ambush Mary.

The hatch used by the KGB to ambush Mary in the dark.

The six houseguests came to
my place for a picnic at the
end of the mission.

CIA Intelligence Star
awarded to me on
May 8, 1980.

Meeting with President Carter
upon my return from Iran
with the six houseguests.

Vice President Bush visits the "Magic Kingdom" in 1986.

Jonna removes her DAGGER disguise in the Oval Office.

The Brandenburg Gate, November 9, 1989, the night the wall came down.

With a Soviet soldier at the Brandenburg Gate on Christmas Eve morning, 1990.

Receiving the Trailblazer Award at the celebration of the CIA's fiftieth anniversary.

More memorabilia of two aliases and one real name.

This headline says it all.

be in a position to threaten blackmail or extortion, not to mention an attempted rape charge.

I countered by inviting her to join David and myself in our suite. Surprisingly, she did so, and even drank a scotch with ice, sitting primly on the edge of the sofa wearing a skimpy wrap. She was indeed an Anglo-Indian, from Calcutta, and said that she had a good Christian upbringing. David and I made a point of praising the tourist marvels we had seen that day in the city and mentioned the sites we planned to visit the next. To this day, I have no idea what Heather's true motive was. She could have been just a lonely young woman looking for company, or she might have had a permanent assignment from SB to probe potentially suspicious foreigners. In any event, David and I kept our cover (and our virtue) intact.

WHEN WE ARRIVED at the cover office the next morning, much better rested, we learned Jacob was coming in from the safe house with Mac. Jacob spent the first hour carefully reading the thick pile of cable traffic that had accumulated since the operation had gone into full swing. Out at the safe site, he hadn't been able to monitor all the messages and had instead relied on Mac for summaries.

"All right," Jacob said, stacking the yellow cable forms in order. "We can't keep twisting in the wind, waiting for Headquarters to pull the stick out." Now he spoke emphatically. "Our young subject simply cannot take much more of this marking time."

Jacob explained how he had spent two late-night soul sessions with NESTOR, fueled by the generous supply of Stolichnaya vodka Mac had supplied from operational stores. It was Jacob's approach to bonding with NESTOR and assessing his true character. Jacob decided that the

Russian combined intelligence and initiative with ambition—an unfortunate combination of characteristics for a Soviet espionage officer.

"He's also Russian to the core," Jacob noted, referring to the almost mystical connection the Russian typically feels with his Motherland. This inextricable bond made it very difficult to convince Russians to abandon their entire life and begin again in exile. NESTOR's situation was even more awkward because of his family ties, people he loved dearly, whose careers would be damaged in the best of cases, or who would suffer worse punishment if the KGB implicated them in his defection.

"The lad is wobbly at the knees, I'm afraid," Jacob confided. "I've had to bring him back from the brink more than once. He and I are going to have to go trunk and tail through the terminal and onto the plane."

Everyone in the office grasped the necessity of Jacob's bold suggestion, even though it potentially undercut the need to maintain deniability. If NESTOR were caught in the airport with Jacob nearby, the Russian might unintentionally compromise him. However, it was an unavoidable risk. Jacob would simply have to run interference through exit formalities while maintaining a certain distance from the subject.

Another vital issue was raised: We would have to reconfirm both NESTOR's and Jacob's airline reservations so they could be certain to appear on the flight manifest. In the tightening procedures after NESTOR's disappearance, the Special Branch had continued to insist that all travelers personally go to their airline office with their tickets and travel documents; no reconfirmation would be accepted by telephone.

But now we realized that, given Jacob's skill and experience in dealing with the narrow bureaucratic mentality of the culture, we might be in a position to take advantage of his presence at the airport and finesse this problem. He would arrive for the flight and insist that no one had told him the necessity of reconfirming his reservation, adding that he knew for certain that there were plenty of seats on board the aircraft—a detail we quickly verified through Raymond's local airport contact. Once Jacob determined that he had convinced the clerks at the airport check-in counter, he would signal NESTOR to approach. On the other hand, even if Jacob was unsuccessful and was bumped from the flight, we would still have time to abort the operation without exposing NESTOR.

Jacob's plan to run interference at the airport demonstrated a very important principle of exfiltration tradecraft: When some minor aspect in the security procedures is uncertain or cannot be included in the document package, always err on the side of omission; act innocent, ignorant, and indignant. Jacob would coach NESTOR on how to stubbornly insist that no one had told him to reconfirm his reservation or had issued him a currency declaration form on arrival. Jacob always advised his subjects accordingly, based on personal experience as a TSD expert who routinely pushed at the boundaries of the opposing bureaucracy. He was out of necessity acquainted with many of the officials at the major airports in the region, and he also knew that the tactic of deflecting blame to "those bloody fools up north" might sit well with the local airport authorities, given regional rivalries. In any event, accepting the face-saving excuse was certainly an easier path for the local officials to follow than challenging an outraged Westerner.

"Time to give Headquarters a *fait accompli*," Jacob announced.

It was clear he had decided to proceed with the operation. Raymond and his team concurred.

Jacob sat down and drafted a complete final operations plan, describing in exquisite detail how we would proceed in the exfiltration of NESTOR. Jacob would act as the escort officer through airport controls and onto the TWA flight to Athens scheduled to depart after midnight on Sunday the twenty-second. NESTOR would be transformed into a German-speaking salesman for a European agricultural equipment firm. We would prepare a fallback identity as an English-speaking tourist, to be used if he had to go to ground again due to unforeseen problems at the airport. But if we had to abort, it was more likely we'd put NESTOR on ice for a long time, and there was no telling what contingency we would then use.

"Better make it right the first time," Jacob said, after editing his draft.

The plan also detailed how we would deploy at the airport. I would be on the terminal rooftop observation deck to determine if our exfiltrees made it aboard the TWA flight, then pass a signal to Raymond by telephone. David would drive one getaway car for Jacob if he needed to flee the airport. We needed another officer from out of the country to drive a second getaway car for NESTOR, because we knew that neither Mac nor Raymond could be seen anywhere near the terminal; both were on the SB's list of local suspected CIA operatives.

The necessity for a third officer worked to our advantage, for he could bring in the airline tickets from Bangkok. This fit the cover story that NESTOR had been called while in Bangkok to proceed to the target city, via the capital, before returning to Europe. The first two coupons on NESTOR's tickets would reflect the date of his "arrival" in the country from Bangkok and the subsequent flight south. After I had made the appropriate markings, I would remove the coupons, so that the red carbon transfers would appear on the unused coupon, as if he'd actually traveled on the tickets. A supposed departure from Bangkok would also provide Jacob with a set of tickets reflecting the same itinerary, so he and NESTOR would be booked on the same TWA flight to Athens.

Because they would have ostensibly arrived on the same flight as the officer from Bangkok, we could use his original international airport arrival cachets as the template for the forged entries I would place in NESTOR's and Jacob's documents. Such a "parallel probe collection operation" would ensure that the actual cachets, inks, and immigration officer's handwriting were valid for that day and flight.

After the final operations plan was filed by IMMEDIATE cable to all the field and Headquarters participants, Jacob gave David and myself final instructions for finishing the documents package. He also made sure he had all the necessary components for NESTOR's disguise and travel ensemble, borrowing my Spotmatic to take photographs of NESTOR in disguise for his alias documents. Just before leaving, Jacob noticed I had a handful of long Cuban Churchill cigars, which I had bought at the duty-free counter in Hong Kong, and he immediately seized a few with enthusiasm.

My curiosity later turned into admiration when I realized that the grand master of the complex game was able to see many moves ahead.

The next morning, Mac brought the film of NESTOR in disguise. After I processed and printed it, I compared the Russian's picture from his national diplomatic identity card to the photographs taken by Jacob. My teammate had done an understated but skillful job. He had adapted an expensive, custom-made wig, ordered earlier for a local case officer, from one of our best Hollywood wig-makers. Since NESTOR had already shaved his head, refitting the wig was easier, but Jacob still had to trim the silk base to fit NESTOR's head contours. The hairpiece was well cut, with a ventilated Vanhorn lace front that conveyed a fashionably conservative European look. The color was dark brown, with salt-and-pepper graying at the temples, a style completely different from the Nordic blond, Mickey Rooney pompadour that NESTOR had sported as a Soviet embassy attaché.

In his alias photos, NESTOR's eyebrows had been darkened, and he had been given a slight shadow beneath the eyes to produce the effect of age bags. Clearly Jacob had carved some soft cotton dental rounds into subtle "plumpers," which he had inserted inside NESTOR's cheeks below the gum lines to distend his youthfully lean Slavic cheeks and enhance the middle-aged sag of his face. This technique helped alter the tone of his German-accented English, a key element of his new persona.

In the full-figure record shots of the disguise, the nicely tailored suit that Mac had contributed, along with expensive "elevator" shoes, created the illusion that NESTOR was three inches taller and much lankier. Jacob had finished the disguise with a pair of 18-karat gold Rodenstock, German-made glasses. The beige tint to the lenses masked the blue color of NESTOR's eyes.

David and I set to work entering the disguise changes in NESTOR's height, age, and coloring onto the bio-data physical-description blocks of his alias documents. I then saw that Jacob had surpassed even his own genius. He had furnished us with other photographs of NESTOR wearing a slightly darker wig cut in a different style, posing in a variety of clothes against diverse backgrounds, in both natural and artificial light. These snapshots provided impressions of a man at successive stages of life. We would use them on documents that had allegedly been issued many years earlier to help support his legend as a Western European.

I printed these shots on different photo papers, which I then "aged" by soaking them in a cup of strong tea before drying them with a steam iron. We then completed the aging process by bending the documents appropriately, working by touch. We knew that the subjective feel of the documents was even more important than the data they contained. An experienced immigration officer was more likely to respond to the "feel" of the passport and the demeanor of the bearer than to the accuracy of the data; a good customs officer made an assessment of travelers by the time they were within twenty feet of the checkpoint.

THE DAY BEFORE the exfiltration was a Saturday. David and I worked into the afternoon, putting the final touches on the documents and reviewing our operations plan checklist. The three others had left the office before midday so that the movements of several foreigners on a weekend would not arouse suspicion. Raymond had explained how to exit the floor by opening a combination lock on the sliding metal grate at the end of the corridor.

With each of us carrying one complete set of NESTOR's exfiltration documents concealed on us, David and I left the office around four and made our way through the poorly lit hall to the metal grate. The combination lock opened easily, and we strode toward the glint of daylight at the end of the hall that marked the staircase.

But we nearly panicked when we reached the doorway to the landing and were confronted by another, unexpected metal grate and combination lock blocking our way out. *Who the hell can I call to reach Raymond with a discreet message?* I thought, unable to breathe for a moment. We had to assume the phone lines to the U.S. Consulate and Raymond's home were tapped. If I called and said we were locked in the building, we were basically lighting a flare for the SB.

Worse, if this building employed any type of efficient watchman, he might spot us fumbling with the lock and call the police. A search would reveal that we each carried a separate set of travel documents bearing NESTOR's disguise photograph. Whatever happened to us, the exfiltration would be in serious trouble.

Leaning close in the faint light, I tried the same combination on this lock. Much to our relief, it sprang open. We groped our way down the stairs, avoiding fresh, blood-red splotches of *pan* juice that the char force had spit out along the route. With Murphy's Law in full effect, we found the doorway into the lobby was also secured with a grate and combination lock.

"Three times for good luck," I muttered to David, entering the same combination. Once more the lock sprang open—a hopeful sign that NESTOR's exfiltration would indeed be a success.

◼

THE EXFILTRATION TEAM converged on the international airport just before midnight on April 22. Jacob and NESTOR arrived in an anonymous Hillman Minx driven by the case officer with whom they'd shared the safe house for the previous week. I came in a car driven by "Pete," the officer from Bangkok who'd flown in two days earlier with the airline tickets. We were trailed by another car that David had driven. Pete and David would stay in the airport parking lot with these two cars, motors running—escape vehicles to be used in the event that Jacob and NESTOR had to leave the terminal quickly.

My position was to be on the observation deck of the terminal rooftop, supposedly waiting for someone on an incoming flight. From there, I could watch Jacob and NESTOR leave the departure gate, cross the tarmac, and board the flight to Athens. When I had confirmed that the plane had departed with or without them on board, I could pass on the appropriate signal, using a wrong-number voice code, to the telephone at Raymond's home, where he and Mac were sweating out the operation.

There was a disciplined logic to all these procedures. No suspected CIA officer who might have aroused Special Branch surveillance was anywhere near the airport that night. Our entire team was from out of town. Moreover, if Raymond or Mac had been under active surveillance, the fact that they were enjoying a late dinner might have served to dampen any SB suspicion that NESTOR was in the city.

But if I telephoned Raymond with the bad news code, announcing that NESTOR and Jacob had bolted, he and Mac would leave his home in separate cars to meet the escape vehicles at predetermined locations, and then try to break the trail.

Checking to see if any surveillance was trailing me, I made a pit stop at the terminal lavatory, which now reeked of carbolic disinfectant that almost masked the stench of the drains. Then I casually inspected the antiquated, railway-style arrivals board and saw that TWA 876 was still due to arrive at one A.M. Having established a plausible reason for being at the airport, I climbed to the observation deck. The flight's arrival time came and went. Moths and termites tumbled incessantly in the spotlights above my head. The smit was rising from the valleys to join the city smog as other airliners landed and taxied up to the terminal to occupy all the free parking slots. It was now almost 1:45, and I could only imagine the scene down in the terminal.

Later, Jacob would describe in detail the sequence of events. He and NESTOR arrived at the terminal curbside just before midnight, and Jacob entered the building first. As instructed, NESTOR waited five minutes, fussing with the strap of his suitcase, which contained a collection of European clothes and personal effects provided by TSD. Then, he also entered the terminal.

Since we had already verified that there were seats available on this flight, Jacob was not worried that TWA would bump them, even though they were not manifested locally, unless there was undue SB pressure to do so. He confidently employed his best British regimental style with the manager at the TWA counter, explaining that he was "damned sorry about the balls-up" concerning the reconfirmation muddle. He suggested that others from his hotel planning to take the same flight had been equally ignorant and managed to persuade the TWA manager to modify the manifest so that passengers who had made reservations from elsewhere in the country could be included.

Watching for Jacob's discreet signal to proceed, NESTOR entered the check-in line. He adopted his new persona perfectly. One of his brilliant ad-libs had been to request a soft-sided leather *portefeuille* to carry his documents and "business" papers, which subtly stamped him as a European. He handed over his passport and tickets nonchalantly as he focused on cutting the tip of the big Cuban cigar Jacob had provided from my Hong Kong stash. Three minutes after arriving at the TWA counter, NESTOR had his boarding pass.

As expected, the first real hurdle came at Customs, where all departing passengers had to have their baggage inspected before it could be turned over to the airline. Jacob cleared the counter ahead of NESTOR. But when the Russian arrived with his suitcase, the inspector, a muscular man in a burgundy turban, flourished a copy of the amended manifest.

"Please remain here, sir," he ordered NESTOR, then disappeared into a back room with the Russian's passport and ticket. NESTOR stood there for an agonizing ten minutes, staring at the closed door. A line of other passengers formed behind him, some grumbling, some sheepishly accepting the quirks of officialdom.

Then the door opened and the inspector emerged, accompanied by a husky European in a sweat-spotted dark suit. Startled, NESTOR chomped on his cigar and tried to swallow. The European was a KGB security officer from the Soviet embassy, a man with whom NESTOR had rubbed shoulders for more than two years. He approached the counter and gazed at NESTOR's face; in response, NESTOR audaciously raised the gnawed cigar and lit it carefully with the gold Dunhill lighter Jane had lent to the operation. He exhaled a cloud of fragrant blue smoke in the direction of his former colleague with debonair calm.

Jacob was anxiously watching this drama from a corner of the echoing terminal, ready to bail out if the Special Branch swarmed from the back room of the customs counter.

The KGB man glared at NESTOR, who met his eyes and drummed his fingers impatiently on the shaft of the cigar. In his new persona, NESTOR was a dark-haired, dark-eyed European more than seven centimeters taller than the Soviet fugitive. The thick-faced man with graying sideburns wore his expensive clothes comfortably, making a point of frequently checking the time on his elegant Seiko wristwatch, then turning to chat with the French couple behind him in the stalled customs line.

Skeptical, the KGB security man checked NESTOR's passport once more against the manifest. Hours seemed to pass before he finally handed the documents back to the inspector, spun on his heel, and returned to the back room. A moment later, NESTOR's boarding pass had received the crucial customs stamp.

IT WAS AFTER two A.M. and the TWA flight from Bangkok still had not landed. I had nervously watched a Swissair DC-8 arrive from Riyadh and a Lufthansa 707 from Frankfurt. An Aeroflot IL-62, belching soot from its engines, landed from Tashkent and lumbered up to the gate directly below. There would be even more KGB gumshoes patrolling the terminal, wary of potential defectors, and it was all I could do to keep myself from pacing up and down. Finally, landing lights pierced the brown haze, and the red-and-white TWA Boeing airliner taxied onto the parking ramp.

After a flurry of baggage handling and refueling, a ground steward-ess in a purple uniform led the passengers out to the flight. By now the smit was so dense that I could barely distinguish Jacob and NESTOR walking with the other tired travelers, maintaining a disciplined dis-tance from each other. The boarding stairs were withdrawn, the engines started, and NESTOR's plane took off for Athens.

Groping my way down the murky staircase to the public phone booth, I felt as if I were carrying a rucksack full of wet sandbags. In an hour, I'd be back at the Southern Paradise, sleeping with the air con-ditioner turned up full blast, but I felt sleep would not come easily be-cause my nerves were so frayed by the ordeal.

I knew I had to send the vital departure signal, and although in reality the old Bakelite phone worked perfectly, I would suffer night-mares for decades about being unable to make the crucial call. I dropped the fat brown coin into the slot and dialed the number: a whir, a click. "Is Suzy there?" I asked.

"No!" Raymond replied, feigning indignation at being awakened in the middle of the night.

I knew we shared the same sense of tired relief and satisfaction. We had just pulled off one of the toughest and most important exfiltration operations in our Agency's history.

DURING THE FIRST seven years I spent in the field, I grew from being an inexperienced kid with dreams of grandeur to a seasoned profes-sional with a firm grasp of serious realities.

Those years shaped the rest of my CIA career and, indeed, my life. I had been an integral player on tough cases in towns such as Vientiane, Bangkok, and Delhi, and my operational assignments took me to even more exotic countries such as Nepal, Pakistan, Burma, Bangladesh, and

Ceylon. Serving with officers at those stations and bases during that first, long overseas tour was probably the most rewarding opportunity I could have ever asked for. At the risk of sounding trite, those men and women were truly a special breed. They often operated in complete isolation from colleagues, sometimes under extremely difficult physical conditions. They had to be resourceful, adaptable, and willing to face discomfort and danger with a sense of humor. They reminded me of those I had grown up with in Nevada—people who could turn the most miserable or insecure situation into a daring adventure.

I received my assignment back to Headquarters in 1974. Although Karen and I knew that it was time for our young family to return to the native country they hardly knew, I realized I was actually going to miss the raw emotions and coveted rewards of life in the field. It was a lifestyle I could not easily abandon.

6 ■ HONOR and GAMBIT

> To subdue the enemy without fighting is the
> acme of skill.
>
> —Sun Tzu

The Blue Ridge Mountains, Maryland, July 1974 ■ Since I had been assigned to an indefinite tour at Headquarters as the Agency's new Chief of Disguise, it was time to plant some permanent roots in America. I was almost thirty-four, Karen thirty-one, and the bland suburbs of northern Virginia that we had left seven years before were no longer a sanctuary from drugs and crime. The energy crisis and riots in the cities had also caused us to question the safety and stability of urban life. We wanted our children to learn how to live closely with nature and be active members of a family, not quasi-boarders at some heavily mortgaged townhouse in Fairfax or McLean.

So Karen and I decided to spend our home leave cutting trees and building a cabin on forty acres of the Blue Ridge mountainside we owned near Harper's Ferry. Choosing to pioneer these wooded hills actually made practical sense. We hoped to finish the cabin by fall, then live there while we worked on building our main house further up the hillside. I would commute to Washington each morning by train. The children would attend a simple country school and, we hoped, pass through the shoals of adolescence shielded from the riskier temptations of the 1970s, which had already disrupted the families of several colleagues we'd known overseas.

In Asia, I'd served a valuable apprenticeship under true operational experts, especially Jacob. Those years had also coincided with dramatically innovative breakthroughs on the technical side of clandestine documents and disguises. Because I had been directly involved in several notably successful operations, including the secret debriefings of PASSAGE in Vientiane and the harrowing exfiltration of NESTOR out of the subcontinent, I'd gained experience and earned rank faster than usual, even in an elite corps like the Clandestine Service. But I knew I would be entering an entirely different theater of operations when I reported to Headquarters in August, and I needed to be physically and mentally prepared.

The Nixon administration and the Agency were in the middle of a major upheaval. The war in Indochina was clearly dragging to a grim conclusion as North Vietnam's Soviet and Chinese allies relentlessly dominated the battlefields with war matériel now that American ground forces and air power had been almost completely withdrawn. The White House, reeling from the Watergate scandal, was near collapse, and, sadly, the Central Intelligence Agency had become ensnared in both these disasters.

Nixon had fired CIA Director Richard Helms, a career officer, in January 1973, for refusing to obey his order that the CIA block the FBI's Watergate investigation. The Director who replaced Helms, James Schlesinger, served only four months before becoming Secretary of Defense. But in those sixteen weeks, he left his brand on the Agency. Distrustful of the Clandestine Service, Schlesinger initiated a house cleaning in which the Monopoly pieces were shaken in a bag and redistributed. Schlesinger had a deep suspicion that an OSS–covert operations "old boy" network wielded too much influence in the Agency, so he bureaucratically distanced the Technical Services Division and its main customer, the Clandestine Service (Directorate of Plans), renaming the DP less euphemistically as the Directorate of Operations (DO). TSD was reorganized to become the Office of Technical Service (OTS) and placed under a newly formed Directorate of Science and Technology.

But Schlesinger's greatest impact on the CIA was the "major purge," as Bill Colby, the Director who replaced him, later described the bloodbath. Seven percent of the CIA staff, most of whom were experienced officers from the Clandestine Service, were fired outright, or pressured to settle for early retirement. Colby, who was, ironically, a decorated World War II OSS veteran and covert action operative par excellence, had served as Schlesinger's trusted confederate throughout this brief but dramatic upheaval. Now he was Director of Central Intelligence and seemed intent on enforcing the unprecedented "reforms" that Schlesinger had unleashed.

The waves left in Schlesinger's wake swept throughout the new OTS. Our bag of alleged "dirty tricks" had been linked in the news media to the Agency's assumed guilt in the Watergate mess. In fact, several former Agency officers, including E. Howard Hunt, were part of the White House's notorious Plumbers unit, and they had received some sanctioned technical support from Agency resources. The most notorious assistance came from an OTS disguise officer, who provided the "ill-fitting red wig," which Hunt had used and may also have worn during the most incriminating incident: the burglary of the Los Angeles office of Pentagon Papers author Daniel Ellsberg's psychiatrist. Even after a year had passed, the dust had still not settled on that episode.

I knew some elements of OTS were undermined by bad morale, bitterness, and a resistance to change that pervaded the hierarchy. That negativity simply did not exist during the camaraderie I had known in my first nine years at the Agency.

It was a volatile time to be returning to Headquarters. I was a field-seasoned technical officer who had begun his career as a humble artist in the Graphics bullpen less than ten years earlier. Now I would be handed the reins of the Disguise operation, a section in transition, and must attempt to transform the way business was done at a time when most prudent bureaucratic heads were retreating into their turtle shells, waiting for the long knives to finish their work.

I was tense and anxious about the future, but I reminded myself that I had worked hard and done well. In my arrogance, I felt certain that there was no other officer in the Agency better qualified to be Chief of Disguise. That sense of confidence was not just a weak attempt at bravado. For example, an operation in South Asia the year before had been one in a series of major intelligence victories, where I had played a key role by introducing improved GAMBIT disguise technology. The American people would probably never hear of it, but my initiatives had raised my status in OTS and, perhaps, throughout the Agency as well. I sometimes relived every detail of those gratifying experiences, knowing I would need as much self-assurance as I could muster for what lay ahead.

A South Asian Capital, Summer 1973 ■ "This guy is going south on us, fast," the Chief of Station, "Simon," explained, leaning over his desk and speaking with a crisp but soft precision that was barely audible above the chugging air conditioners. "It's up to you to save the operation. If he doesn't buy the disguise option, he'll surely quit."

The blinds were tightly drawn against the mid-afternoon desert glare. On Simon's desk lay the SECRET case file of his most valuable agent, a senior official in the local government, code-named "HONOR." I studied the assortment of news clippings and surveillance photos of the agent, a portly, distinguished, middle-aged man wearing the well-tailored flannels of a British academic. Since independence, HONOR had risen to the high ranks of his government, with assignments in several European and Middle Eastern cities.

By late summer 1973, HONOR had assumed a vital national security position in his capital. Because of family and political connections, he was highly respected by his peers and superiors, none of whom suspected he had also been a well-regarded CIA agent, providing vital intelligence for years. We considered him a long-term, high-level, unilateral penetration into the host government, privy to its most sensitive military and political secrets. But he had mostly been posted overseas throughout his clandestine relationship with us.

In many ways, the main problem in HONOR's case was similar to the difficulties we had overcome with PASSAGE in Vientiane. While HONOR had served in his country's embassies, his debriefings had taken place in lavish private rooms of elite clubs and restaurants in London or Paris whenever he could slip away for a brief holiday. In those encounters, he had felt secure enough to pass over extremely valuable diplomatic cables and dispatches, and to discuss in detail the nuances of the constantly shifting alliances within the Soviet-dominated, "nonaligned" group of nations, of which his country was one of the unofficial military leaders.

In other words, HONOR was not just another "fat cat" Asian official, but in fact possessed vital insight into the heart of political and military relationships that radiated out from South Asia into China, throughout the Middle East, and on to the Soviet Union. Because we had paid him well, he had made it a point to ferret out all the sensitive information that the Langley analysts presented on their shopping lists. Consistently that information had been accurate.

Then, much to our dismay, HONOR had been kicked upstairs to become a senior national security adviser to the cabinet. He could no longer meet with his case officer in a discreet back room in a St. James club, or a private dining alcove of a Quai d'Orsay restaurant. Now he had to conduct weekly clandestine rendezvous in this dusty little capital, where local security was ubiquitous and jealous unofficial spies in his own ministry dreamed of seeing him hurled from his pedestal.

"His feet aren't just cold," Simon noted. "They're frozen solid."

I flipped through Simon's latest debriefing notes in the 201 file. HONOR was terrified about getting caught here in his own country because he clearly recognized the consequences: Not only would he be brutally interrogated, then most likely executed, but his family might face the same fate. Even if they survived, the considerable "unofficial" fortune he had accumulated would be confiscated.

Because his intelligence value to the United States had only increased since his return from overseas, veteran case officers like Simon were not about to let him slip away without a fight. HONOR's country was actively negotiating with the Chinese on a highly sensitive and secret arrangement that would allow a few Chinese intermediate-range nuclear missiles to be stationed in the country. This arrangement would counterbalance a similar Russian rocket threat, which a neighboring nation was suspected of developing.

But there was more. HONOR's national air force provided unofficial mercenaries throughout the Middle East who were especially active in Egypt. As defense adviser, HONOR had begun to hear ominous reports that the Soviets had moved tactical nuclear weapons into Egypt to be used against the Israelis, in the event of yet another Arab-Israeli war.

HONOR's frostbitten feet, however, had not moved him to the last scheduled clandestine meeting, nor to the alternate meeting. He was obviously on the brink of "heading south," as Simon put it, just when American intelligence needed him more than ever. If tensions in the Middle East again reached crisis level, America was going to need every reliable intelligence source with knowledge of the region it could possibly find. Simon simply could not afford to lose HONOR now.

I returned the thick case file to Simon's desk. Given the gravity of the situation, I had to find a way to convince HONOR that we could provide a disguise so foolproof that he could remain an active member of the operation and work virtually under the nose of the local security service in this claustrophobic little city.

"Got any good ideas how you're going to bring this guy along?" Simon asked skeptically.

Simon was what we techs called a "James Bond" case officer; when he went operational, he favored black turtlenecks, Italian driving gloves, and expensive Harris tweed. His graying brown hair was cut in the collar-length British style and, whenever he took off his sportscoat, I half expected to see a Walther PPK slung in a shoulder holster. Although his sleek appearance might attract too much attention on a surveillance detection route through narrow back streets, he was a highly competent officer.

"I do have a couple of ideas, Simon," I said, suppressing a smile. "First, the best way to convince HONOR to use the new GAMBIT disguise is to actually deceive him with it, to put him in the place of his own security service and show him how well it works."

"Let's give it a go," Simon said.

THE CRUCIAL TEST came just after sunset two days later. I strolled through the cool, deepening twilight under the jacaranda trees surrounding walled diplomatic and expatriate villas. It was certainly not unusual for a Westerner to be savoring the early evening breeze after enduring the arid heat of the late summer day. As I crunched up the gravel path of the well-maintained park and entered the murkier shadows beneath the acacia trees, I knew I had been walking for some time with no one behind me.

It was ominously black under the branches as I groped my way to the stucco wall of the villa at the far end of the park. Reaching out in the darkness, I immediately touched the slim package that "Carol" had placed there from her garden on the other side of the wall. She was a seasoned case officer used to operating under an innocuous cover in the backwaters of the world, and she managed to maintain a convincing legend at this tasteful villa, where she often entertained diplomats and local officials.

I followed a circuitous route toward the glow of the streetlamps beyond the trees on the other side of the park. Satisfied that there was no one nearby, I slipped the GAMBIT from the black bag and applied the disguise, working easily by touch alone in the shadows, just as I had done on so many practice runs in my darkened hotel room. I checked each prominent part of the disguise and completed the look with a colorful cravat, tucking it in by running my right finger around the collar of my dark cotton shirt. Satisfied that my transformation was complete and that looked perfectly natural, I strolled along the gravel path on the opposite side of the park from which I had entered.

Now back on the sidewalk, I passed the occasional lone figure or couple walking in the pleasant evening coolness. People nodded politely, and one man even offered a cordial greeting in English: "A very good evening to you, sir."

I returned his nod but dared not risk a response, which could have been distorted by the disguise.

The couple passed, and I breathed easier. Thankfully, the improvements that Jerome Calloway and the TSD technicians had made in the GAMBIT disguise, based on suggestions that I had forwarded through channels, were more effective than any of us had hoped. As I continued strolling, my encounters with local people beneath the glaring mercury-vapor lamps illuminating the street were relaxed and pleasant. In spite of the exercise, I wasn't even sweating. *This is going to be a piece of cake,* I thought after a fifteen-minute trial run through the neighborhood.

Then, I saw a Westerner in wrinkled tennis whites walking a huge black Belgian shepherd coming straight at me down the sidewalk. The dog was a massive creature, tall and barrel-chested, with a head like a wolf's. This beast was, without a doubt, a very effective watchdog. But how the hell would he react to me? Despite Jacob's constant warnings to think of every possible contingency, none of us had even considered this particular development. GAMBIT worked well in both natural and artificial light, but we never asked ourselves how the disguise might look to a fierce black dog, marching along with nothing but a choke collar restraining him.

I could bolt across the street, of course. But that was not normal behavior for a Westerner, especially one whom GAMBIT had transformed into an almost stereotypical, pink-faced Brit expat, who should by definition be a dog lover. So I held my ground on the sidewalk, trying to remember everything I had been taught as a kid back in Nevada about animals being able to sense human fear. As a ten-year-old in Caliente, I'd always followed those lessons whenever I encountered menacing strays on the street: Look 'em in the eye, keep your shoulders squared, and walk right past them.

I swallowed hard and continued strolling casually, trying not to betray my apprehension. I pictured this unpredictable monster wheeling suddenly, its jaws snapping to attack, but a moment later, the dog and its owner passed close by without incident.

Trying to relax I approached the turn at the front gate of Carol's house. My next challenge was to walk past her *chokadar*, who was squatting in the middle of her open gateway with three of his friends, enjoying a few crumbs of hashish in a clay pipe, which they passed among themselves in the crisp night air. The watchman looked up and nodded with a snaggle-toothed grin, offering a serene *salaam*.

As I stepped inside Carol's gate, I made a mental note that we had just achieved a professional milestone. On two other operations, we had used these new disguises at night on individuals riding in vehicles—the PASSAGE debriefings in Vientiane, and a series of equally productive agent-case officer meetings in a neighboring South Asian country. These latter debriefings were of military officers with detailed knowledge of newly delivered Soviet weapons, including T-62 tanks, the latest variant of MiG-21 fighters, and, most important, the new upgrades to SAM-2 antiaircraft missiles. American pilots in Indochina and our South Vietnamese allies were confronting these same weapons on the battlefield, and the knowledge we gained about their technical properties helped defeat the enemy during the Communist Easter Offensive of 1972 and the bombing of Hanoi in December. All of the agents wearing the GAM-BIT disguises had passed the scrutiny of suspicious local police and even trained counterintelligence officers as they rode in our cars disguised as Westerners. However, we had not considered GAMBIT at that point in its evolution to be practical for the type of operational environment HONOR faced.

In other words, if an agent or case officer needed hours of pains-taking preparation at the hands of an expert to pass close scrutiny in ordinary light, then routine use of disguise was not practical. And, be-fore our meetings, we would not have the luxury of sitting down with HONOR for such protracted periods of time because his new position made him inaccessible. For this reason, the success of tonight's "moist" run, dead-drop pickup of the disguise was extremely crucial. Dead drops were quick, safe, and tested elemental tradecraft; they could be con-stantly changed, and, if chosen well, were hard to detect. Dead drops usually served as clandestine mailboxes for messages, but working with Calloway and TSD, I hoped to perfect a GAMBIT disguise that could be folded neatly into a flat black bag and applied in the dark in a few minutes without adhesive, mirrors, or expert hands. My evening prom-enade around the well-lit streets of this prosperous neighborhood, in-cluding the unexpected encounter with the watchdog, proved that our increasingly paranoid but valuable agent HONOR could be taught to disguise himself with a GAMBIT he retrieved from a dead drop.

This strategy was an essential part of making our operation con-vincing, secure, and feasible. Obviously, an agent who had already grown nervous would never even consider hiding something as incrim-inating as the GAMBIT disguise at his home, where its discovery would not only compromise his own safety but that of his entire family. Nor would we want to leave such an obvious piece of spy gear in his care.

As I walked up Carol's winding drive between hibiscus hedges, she came down the front steps to greet me, smiling as if I were an old friend. The *chokadar*s squatting over their pipe were not particularly interested in the distinguished Western gentleman calling on the *memsahib* after dinner. She often hosted late evening bridge games during the warm months, and a new partner aroused no suspicion. It was an excellent cover because we were already planning for HONOR to become a true connoisseur of the game.

Once we were inside, I faced Carol. "What do you think?"

I purposely avoided using the words "look" or "disguise," a habit we all followed out of discipline, assuming as we always did that any location might be bugged.

"Fabulous," she said, shaking her head in wonder. She peered more closely, her smile widening. "I never would have guessed it." We went into her living room, where Bokhara carpets were spread tastefully on the muted tan tiled floor.

"Have a seat." Carol pointed to a handsome leather chair in a corner. "Simon should be here shortly."

I sat down, crossed my legs comfortably, and inspected the cigarette case and ashtray on the copper table beside the chair. "When he comes in with his friend, I'll be sitting here smoking a cigarette and drinking a gin and tonic, just as we discussed," I murmured to her under my breath.

When she came back from the kitchen with the drink, I placed it on the table, then reached up to adjust the angle of the lamp on the other side of the chair. The room was dotted with pools of light from similar lamps. Carol stood there watching as I lit my cigarette and sat back in my chair.

"Still okay?" I asked again.

"Smashing," Carol said with a giggle. She disappeared into her bedroom, where she would remain until our visitors were gone.

I had just lit my second Dunhill 100 when Simon and HONOR arrived. Simon ushered the agent into the living room and asked him to take a seat diagonally across the room from me. The dapper, plump gentleman was clearly flustered from his ride across the city, crouching on the floorboards of Simon's ops vehicle (his personal, English racing-green sports car). Simon had picked him up in an obscure alley, and HONOR had succumbed to this nerve-racking indignity in order to avoid being spotted in the company of the American "diplomat" widely rumored to belong to the CIA. I saw immediately that Simon had been right: Here, indeed, was an agent on the verge of quitting.

Tonight, Simon was again playing the role of the glamorous Ian Fleming spymaster. He had, most likely, terrified our agitated agent, who squatted suffocating in the narrow space in front of the passenger seat as Simon careened through traffic with his usual flamboyance.

Turning to leave the room, Simon smiled benevolently at HONOR. "I'll be right back."

But, as planned, Simon did not introduce me or even acknowledge my presence before he left. HONOR, an old-school diplomat who observed a rigid social code, was left stranded, sitting in a room twenty feet away from a complete stranger, who silently smoked and drank a gin and tonic without even glancing in his direction.

Simon returned, leaned over HONOR's chair, and spoke quietly before turning toward me.

"Arthur," Simon said, using HONOR's pseudonym, "I would like to introduce Dr. Anderson from Washington, who came here to meet with you tonight."

HONOR rose somewhat nervously as I approached him.

"It's a great pleasure to meet you, sir," I said respectfully as I shook his hand.

"It's my pleasure . . ." HONOR never finished his deferential responses. He gasped as he watched me bring my hands up to my face, and, within seconds, the GAMBIT persona evaporated as I emerged like a butterfly from its chrysalis. He recoiled with a mixture of shock and delight, recognizing me immediately from an earlier meeting. Although he had witnessed what had happened with his own eyes, it was clear that he could not fully comprehend how we had managed to pull off such an incredible act of sorcery.

"Amazing," he finally muttered. "Simply incredible."

No more discussion was needed. We followed Simon into another room, where I set to work on the fitting of the GAMBIT disguise designed expressly for HONOR. The training in its application and removal took less than half an hour. He was so delighted with the realism of the disguise, however, that he spent a whole five minutes in front of the mirror, experimenting with different facial expressions and launching into an impressive imitation of the corrupt prime minister for whom he worked.

I then showed HONOR the diagrams of the areas in the park where Carol would drop the GAMBIT twenty minutes before scheduled meetings. We asked him to memorize street maps of the routes he would take to and from the park, and then on to Carol's home for "bridge games."

My GAMBIT training work with HONOR ended that night. Simon, Carol, and their highly valued agent continued with their mission. From that point on, HONOR voiced no qualms about meeting Simon for extended face-to-face debriefings on vital intelligence matters.

As we had suspected, HONOR's intelligence work became essential that October when Israel and its Arab neighbors collided in the tragic bloodbath known as the Yom Kippur War. For more than two weeks, the Israelis fought hammer-and-tongs with the Egyptians and Syrians, with unprecedented losses of men and equipment on all sides. Threatened with defeat early in the conflict, Israel put its small nuclear missile force on alert. With their respective allies crumbling, both the Soviet Union and the United States conducted massive sea and air lifts to resupply the opposing sides.

But intelligence reports from all sources reaching Headquarters, including those from Soviet bloc stations, indicated that the Soviets were doing more than just replenishing the supply of MiGs, tanks, and SAM missiles that their Egyptian surrogates had lost in combat. There were signs that Soviet leader Leonid Brezhnev planned to counter the defensive Israeli nuclear alert by sending Strategic Rocket Forces teams, equipped with nuclear warheads, to support the Egyptians' Scud missiles, already aimed at Israel. Once again, the Agency issued an all-points alert to stations and bases, with urgent requests for more data.

I happened to be back in the same dusty Asian capital when Headquarters' message reached the Station. That afternoon, Simon sent HONOR an emergency meeting signal: a small "X" hastily scrawled in charcoal on a gaudy cinema poster, plastered on a mud-brick wall near a kebab stand. HONOR went to the suburban park that night, found the GAMBIT in its drop site, and emerged from the shadows, a Westerner sauntering confidently through the chilly autumn evening to join an impromptu bridge game at Carol's villa.

As I adjusted and repaired HONOR's disguise, which showed signs of intensive use, Simon relayed the Agency's urgent request for information.

"We have to know if the Soviets have already shipped nuclear weapons to Egypt," Simon said, his voice strained.

"Oh, they have," HONOR replied casually as if he were discussing the price of lamb in the bazaar.

"How do you know that?" Simon demanded.

HONOR calmly recounted that his office had received an intelligence report a few days earlier from one of its military officers assigned to the Egyptian forces along the Suez Canal. As the officer deployed to a forward base, he encountered small clusters of Soviet troops at Egyptian missile sites and concluded these Russian soldiers were guarding camouflaged concrete bunkers containing nuclear warheads for the Scuds.

Simon and I exchanged a tense glance. HONOR's information was shocking. If the fragile cease-fire on the battlefield was broken and heavy fighting resumed, the war could quickly surpass the nuclear threshold. It seemed that World War III, or at least something dangerously close to it, was about to explode in the wastelands of the Sinai and Negev deserts.

Simon quickly left Carol's house and sent a FLASH precedence intel cable, attention DIRECTOR. After integrating HONOR's report with other intelligence indicators, the CIA urgently informed National Security Adviser Henry Kissinger, who advised President Nixon to immediately order American military forces to DEFCON 3, a high state of nuclear alert. It was a long night for the National Security team in the White House situation room. DCI Colby was in constant communication with Langley. American Polaris submarines refined the targeting data of their ballistic missiles, while our lumbering B-52 bombers held their orbits in the Arctic and Mediterranean, awaiting orders to proceed toward Soviet airspace. The Middle East powder keg seemed ready to explode. Indeed, the global security situation had not been so precarious since the Cuban missile crisis.

But the Soviets backed down. They canceled the orders of their airborne troops preparing to fly to the Middle East, withdrew the nuclear warheads already positioned at the Egyptian Scud sites, and recalled their ships. Two days later, when the White House was convinced of the Soviets' new intention, Nixon rescinded DEFCON 3. In the ensuing confusion, Nixon's many critics accused him of orchestrating this crisis to divert the nation's attention from the heightened Senate Watergate investigation.

I did not learn all the inside details of this frightening confrontation until late 1974, when I attended the Agency's Mid-Career course for promising officers of middle grade. My roommate in the course, "Bob," was from the Directorate of Intelligence and had been the "duty" analyst on the line with DCI Colby during the tense days of the confrontation. Bob related to the class that Colby had been installed in the White House Situation Room, as all signs indicated that Brezhnev had sent warheads to Egypt and that the United States and the Soviet Union stood face-to-face, with their nuclear weapons cocked and ready to fire.

"It seems we were *that* close to a thermonuclear war," Bob told us. "But the Sovs knew that we were onto their nukes in Egypt, and we weren't bluffing. They had no choice but to back off."

I was secretly proud of what my colleagues and I had accomplished, and I was more determined than ever to enhance our operational capability, as long as the Agency supported my efforts. I felt this resolve in the face of weakened morale and a strong sense within the ranks of the CIA that we had to fend off attacks not only from our Soviet bloc opposition but from many in our own government and the public. We were in a state of siege.

DESPITE THE GLARE of unflattering public scrutiny in the mid-1970s, important members of OTS and the Agency considered the HONOR case to be a breakthrough in the evolution of the GAMBIT disguise, partly because it had served to convince a vital agent that we would spare no expense and explore technical frontiers in order to protect those like him. In addition, we soon found that there was a marked increase in the need for enhanced disguise technology in other third world capitals as we struggled to keep up with the rapid spread of Communist influence in Asia, Africa, and Latin America.

Because of operations like the HONOR case, I was considered the de facto expert in disguise operations, which called for the invention of startling new techniques and materials, as well as the subtle perspective and touch of the artist. Even before I left Bangkok, I was keeping Headquarters apprised of how improved disguise and related tradecraft might be employed in other areas so that our case officers could conduct personal meetings with agents. Such a possibility was especially important to explore in cases where agents brought in cumbersome Soviet weapons manuals, which were most efficiently photographed with large fixed cameras best used in more secure facilities.

The traditional East-West geopolitical issues of the Cold War were merging with other global problems, such as terrorism and international narcotics trafficking. Our case officers might find themselves debriefing a Soviet GRU agent one day and a source with information on the German terrorist Bader-Meinhof gang the next. As had been the case with HONOR, all agents vacillated, fearful about the risk of compromise. Ultimately, we hoped, we could make disguises so realistic, comfortable, and easy to use that case officers worldwide would automatically turn to us for our services. I recognized, however, that this vision was very ambitious.

7 ▪ Pinball

If you ain't Audio, you ain't shit . . .
—"Schraider"

South Building, August 1974 ▪ "We want to make the disguise capability of the Office of Technical Service the best it can be, Tony. Absolutely second to none."

"Big Arthur," the Chief of Operations, leaned across his desk on the second floor of the South Building, while my immediate boss, "Tim," the Chief of Authentication, and I absorbed this Olympian pronouncement. After all, the Soviet bloc opposition had about a forty-year head-start on us. But the Agency believed that the KGB and its allied services had nothing even approaching our inventive GAMBIT disguises. Since Big Arthur, Tim, and I had all just returned from overseas assignments and were new to our current positions, these lofty ambitions seemed appropriate, despite of the shaky morale and anxiety we felt all around us.

The mood in the Agency was uncertain. The previous afternoon, a few of the employees had trooped out to the Mall to watch *Marine One*, President Nixon's helicopter, clatter off the south lawn of the White House en route to Andrews Air Force Base, in his last official flight as Commander-in-Chief. Some felt that with Nixon's resignation, the Executive Branch was destined for hard times. He had dragged the Agency into the Watergate scandal, and many ambitious people in Congress were eager to score political points by kicking us as we reeled from the media's blows. I immediately noticed that some of the old Far Eastern hands, catching an after-work drink in O'Toole's, no longer had their CIA ID cards on neck chains tucked into their shirt pockets—once a brash statement that they were players on a top-ranked team. They were losing the esprit de corps that had always bolstered their commitment to the relatively low pay and bone-crushing work schedule underpinning the James Bond fantasy world of espionage. Of course, this negativity could have been attributed just as easily to the disaster in Vietnam.

One member of the disguise group had been directly entangled in the Watergate investigation, so keeping morale high in my section was another potential challenge I had never faced before. Since I was now in charge of what many saw as the most unpredictable form of operational tradecraft, I was in a delicate position; however, I reminded myself that my superiors had given me the go-ahead to press on, and I intended to do so.

One of my first assignments was to review a list of "questionable activities" that new DCI Bill Colby was preparing for the Senate investigation of CIA. This list included sensitive secrets dating back to the Agency's birth in 1947. Soon to be known as the CIA's "family jewels," these secrets were subjected to even wider scrutiny by congressional committees and the press. The Agency's Inspector General was already hard at work collecting data on such delicate subjects as Agency-sponsored assassination attempts against Fidel Castro and leftist Congolese premier Patrice Lumumba, among others.

With the Ford White House distancing itself from the CIA, Colby was in a lonely position. The CIA's blood was in the water. The *New York Times* investigative reporter Seymour Hersh discovered that the Agency had helped run counterintelligence operations against certain American antiwar groups in the 1960s and early 1970s on U.S. soil, in violation of the CIA's charter forbidding domestic spying on American citizens. Hersh wanted to interview Colby about these operations—known as CHAOS—and Colby agreed.

Begun in 1967, CHAOS had been conducted, in cooperation with the FBI, on the direct orders of President Johnson and continued under Nixon. It had been based on a valid premise but was of questionable legality: The Agency had reliable information that Soviet bloc intelligence services were lending material support to a few of the most radical antiwar groups protesting our involvement in Vietnam. But the FBI, which had a legitimate charter to conduct such a counterintelligence operation in the U.S., did not have access to the CIA's sources. Therefore, it made sense for the two agencies to cooperate to meet the White House requirements.

The Hersh story on CHAOS was on the *Times* front page on December 22, 1974, under the headline HUGE CIA OPERATION REPORTED IN U.S. AGAINST ANTI-WAR FORCES, OTHER DISSIDENTS IN NIXON YEARS. The article described CHAOS as a "massive illegal domestic intelligence operation." It was technically illegal, but it was not massive: Just over seven thousand out of the millions of American antiwar protesters had been briefly targeted, then cleared. Of these, several hundred suspected of channeling East bloc resources into the most radical groups operating in the United States were placed under more intensive surveillance. As in any counterintelligence operation, the CIA-FBI methods appeared unsavory, including infiltration of agent informers, mail intercepts, electronic eavesdropping, and street surveillance.

William Colby noted in his memoir, *Honorable Men*, that the Hersh article "raised the specter of a government agency running amok, becoming a Gestapo, violating the fundamental constitutional rights of the American people." He added that the article was just one blow in a long string of accusations since 1947 that brought the CIA to the depths of despair. "Politicians, editorialists and ordinary citizens demanded an end to the CIA's heinous practices. Devastating charges were hurled; the Agency was termed a 'rogue elephant' out of control, a threat to the nation's fundamental liberties, a Frankenstein monster that had to be destroyed."

Such was the prevailing atmosphere when I plunged into my Headquarters assignment in 1974.

By its very name, the new Directorate of Science and Technology signified a not-too-subtle shift away from the traditional practices of espionage involving case officers and agents—Human Intelligence (HUMINT)—toward a much greater reliance on technical methods of intelligence gathering, many of them involving some form of electronic eavesdropping by satellite, "overhead" imaging with spy planes and satellites, and over-the-horizon radars tracking Soviet bloc military aircraft. How would the small elite team of Clandestine Service mavericks fare in this arcane environment?

For the Agency, there was a definite political advantage to shift to technology. From the media's perspective, electronic equipment was less malevolent than "renegade" officers: Satellites did their work without the potential embarrassment of human espionage.

The growing emphasis on technical collection, combined with the mounting national scorn for covert operations, meant some audacious and resourceful case officers might become too cautious. Such an escalating state of paralysis and loss of HUMINT would have been catastrophic at that crucial point in the Cold War. We were reeling from the effects of Watergate, the loss of the war in Indochina, and blatant Soviet attempts to subvert legitimate postcolonial struggles. Soviet bloc intelligence services and military missions had never been so active on a worldwide scale. At the same time, the conventional and nuclear warfighting capability of the Warsaw Pact was being upgraded, while the KGB was running scores of black propaganda and subversion operations designed to undermine the NATO alliance. It was not just a coincidence that well-funded and well-organized urban terrorist groups flourished in Western Europe at the time.

The United States and its allies desperately needed the skills and resources of the CIA's Clandestine Service, and I was determined to do my part. Riding the little blue bus back from Langley to Foggy Bottom one day, I suddenly recalled a mantra my mother had taught me as a child when times were rough. "Focus on the task at hand, be of good cheer, and things will sort themselves out."

WHEN I GOT ready to work in OTS's disguise facility, which occupied a suite of four small rooms on the third floor of the Central Building, I sensed that some of the office's senior managers held a cynical view about disguise tradecraft, even though most were interested in expanding the practice.

One manager summarized the prevailing attitude among veteran case officers, especially those who had earned their rank recruiting and running agents in European cities and westernized capitals elsewhere. "Even those who have fairly effective disguises say they really don't need them and almost never use them," this official told me. These officers had matured professionally, he explained, when disguises amounted to little more than blatantly phony wigs, mustaches, and dark glasses. These experienced "first team" officers considered a disguise an amateurish crutch. But a lot of them hadn't worked in Eastern Europe or in third world cities, where the allies of Soviet bloc security services blanketed the ground with surveillance networks.

In South Asia, for example, the security services could use the incredibly rich diversity of street life to their advantage. I once saw a photograph of one service's annual disguise competition from this period, which I studied with amusement and admiration. There were scores of men, women, and even children garbed in outlandish outfits: water wallahs with their bulging goatskin bags and tinkling brass cups, snake charmers with billowing pantaloons, elephant *mahouts*, pseudo-leprous beggars, and *harijan* (untouchable) sweepers. These individuals would have been invisible to everyone, including old hands among the station officers. Vigilant surveillance had already compromised several of our intelligence networks, and just the constant threat of surveillance served as a deterrent to citizens who might have been tempted to sell interesting intelligence to the CIA. Therefore, creating robust and flexible disguises for both station officers and their assets was in fact a mounting and unmet need.

But other case officers, the official added, especially station chiefs under pressure to recruit more assets using fewer resources, felt OTS was "pushing" disguises instead of answering their need for other "hard" technical devices and countersurveillance equipment.

I felt strongly it was very important to have disguises prepared, ready to use, even if the officers' immediate situation did not require them. We could never predict when an officer or an asset might suddenly need to pass close scrutiny undetected. But the prevailing philosophy was that officers should wait until they arrived in the field to assess their disguise requirements, then send for an OTS specialist if necessary. Therefore, one of my priorities was to educate as many case officers as I could about all types of disguise techniques, both traditional and innovative, as well as the tradecraft necessary for their optimal use.

"We're going to have to start out at the Farm," I told Big Arthur one snowy afternoon in December. "The name of the game is planning for the future. The Agency's future obviously lies with the Career Trainees taking the Operations Course at Camp Swampy. This is the group we have to inspire."

Big Arthur squinted reflectively. He understood my logic: If we could overcome the initial psychological resistance to disguises early in an officer's career, these newly minted professionals would become our best clients when they reached the field.

We both realized, however, that this aversion to disguise was deeply seated in us all. The public still thinks that American case officers and foreign spies gleefully pull on "Mission Impossible" masks three or four times a day to trick the opposition. The reality was exactly the opposite. Most of my colleagues dreaded the idea of drastically altering their appearances, and occasionally even their sex, as part of their tradecraft training. I knew the best way to overcome this hesitation was through peer influence, cultivated in the structured environment and classes of the Farm's Operations Course. I envisioned a workshop where all the trainees received friendly but effective disguise instruction from OTS experts, delivered in an atmosphere in which each disguise could be critiqued and improved without the intimidating pressures of the pecking order that prevailed among the old boys at Langley.

That winter I met with the DO's top training officers on the fourth floor of Headquarters.

"There's a real gap in disguise tradecraft training out at the Farm," the Operational Training chief conceded, "and more of our operations and agents are being rolled up in the so-called 'benign' operating environments abroad."

His superior dragged the informal ledger he always kept nearby to the center of his desk. As in any bureaucracy, The Budget was sacrosanct. "Let's begin by sneaking in a line item for FY 76 and see if anybody dares to shoot it down."

With that straightforward discussion, a wedge had been placed in the door of DO tradition. Eventually, after several years of dealing with Headquarters politics, I saw my goal met: All new case officers learned disguise techniques and their related tradecraft at the Farm, and already had a second- or third-generation disguise by the time they went to the field.

I OWE MUCH of the success of this effort to one of my Agency mentors, a battle-hardened case officer whom I'll call "Bull Monahan." When I'd first met Bull in the Far East, he had the reputation of being a "technophile" who totally immersed himself in any new operational gadget or piece of hardware that TSD produced. For example, during a European tour in the early days of the Cold War, Bull had become fascinated with microdot agent communications, and he soon was one of the Agency's most successful practitioners of this esoteric technique. In the

Far East, he became engrossed with locks and picks and ran a number of audacious "quick plant" audio operations against hard Communist targets, such as the hotel rooms and offices of Soviet bloc diplomats.

He was in his lock-picking phase when I met him as a young technical officer newly arrived from the States. I entered his office one morning to find him hunched over his desk, a jeweler's loupe on one eye and his nose almost touching the gleaming steel clockwork of a complex tumbler lock. He didn't even look up to greet me and merely answered my questions with his trademark grunts.

Six years later, while I was experimenting with GAMBIT disguises in Bangkok, I encountered Bull Monahan again. He immediately saw the potential of these improved disguises, and I did some experimental work for him and his case officers. Bull's fascination with the "gee-whiz" aspects of espionage technology did not blind him to the complexities of what he called the "operational stage": all of the elements impinging on the problem at hand, such as locale, level of surveillance, and people involved.

"A disguise is only a tool, Tony," he told me. "Before you use any tradecraft tool, you have to set up the opposition for the deception."

I was reminded of the magic tricks I'd performed on those long winter nights around our wood stove in the parlor of the old house in Caliente. I had learned then a powerful lesson: You had to misdirect the attention and interest of the audience to execute a believable illusion.

In the Far East, for example, Bull had enjoyed great success running both local and Soviet bloc agents, even though he was often under surveillance by the best of the local services and the KGB. He went to a provincial tailor who sewed him a reversible raincoat, beige on one side, dark gray on the other. Dressed in the coat and a tweed cap, lugging a bulky shopping bag, Bull would travel around the center of the city on the subway. Then he would duck into carefully selected cul de sacs, reverse the coat, pull on a beret and glasses, and deflate the balloons inside the shopping bag that gave the illusion of weighty solidity. Less than twenty seconds later, he'd reappear in the subway corridor, limping with the aid of a collapsible cane, and watch surveillance stream by.

Only when he was certain that he had shaken his tail would he continue on his path to an agent meeting.

I was fascinated to learn that Bull Monahan employed an equally elaborate degree of preparation when he went back to Headquarters on shopping expeditions. From his years as a successful case officer and senior manager of case officers, Monahan knew that the man in the field could best seize the attention of the lethargic giant that was Headquarters by peppering Langley with a barrage of carefully crafted "rockets"— cables designed to attract the attention of the appropriate powers and incite them to action.

Monahan's rockets were always marked high priority, usually IM-MEDIATE, and sometimes even NIACT. They were rarely lost in the Agency's daily avalanche of incoming paper. Monahan unleashed this barrage in the weeks before boarding the plane for Headquarters consultations, ensuring that he'd have a number of urgent meetings lined up to address the items on his agenda. That way, he could put aside more mundane business, such as budget reviews and other administrative drudgery.

In a sense, Monahan was throwing himself into the middle of the playing field like a runaway kick-off returner, forcing the opposition to chase him. He'd practiced this routine for years and it was usually successful, partially because he worked important cases and needed assistance, whether it meant budgetary support, more personnel, or stronger technical aid on an urgent basis. Monahan's genius was coupling the power of networking with the occasional grandstand play. With hard work and strokes of exquisitely timed good luck, he was a remarkable figure to observe.

As I watched him consistently acquire the most-qualified case officers, rare financial assistance, and advanced OTS gadgetry for his stations, I realized Bull Monahan was a grand master at a Headquarters game I privately came to call Pinball.

The object of Pinball was to place the ball (your idea) on the table (the operational venue inside or outside the corridors of power) and keep it there as long as possible to see how high a score you could rack up. Although Monahan worked in the serious world of espionage, he taught me that back at Headquarters, competing for budget, staff, and technical resources was, in fact, a game. If you took it too seriously, you'd tighten up and lose.

East Building, August 1975 ■ "We need to put three round-eyes into that building along with the local landlord . . . in broad goddamned daylight," "Schraider" said, glaring malevolently across his desk at me in the manner that had earned him the nickname "Darth" among his elite cadre of audio penetration officers.

Schraider was the Chief of OTS Audio Operations and was another Agency legend. The fact that I had been granted the privilege of a personal meeting, rather than being issued a requirement memo, told me that Schraider had an important operation planned, and that he'd also heard positive news about the Disguise program and wanted to test my mettle. He sat behind a massive walnut desk that might have belonged to Allen Dulles when he'd occupied this same suite as DCI. Schraider's cold blue eyes were probing, and he slapped the desk with a swagger stick for added emphasis as he spoke.

On the dark paneled wall behind him, Schraider had displayed his trophies of the shadow wars: an AK-47 mounted above a menacing arrangement of exotic weaponry, including a Meo flintlock, a Yao crossbow, a glinting Ghurka "ball-cutter" knife, and a Pygmy spear from the Congo's Ituri Forest. At the center of the desk, facing any visitor of the Chief, sat a brass placard on which was boldly engraved IF YOU AIN'T AUDIO, YOU AIN'T SHIT.

That statement summarized Schraider's philosophy. For him, audio penetration was the epitome of espionage. Infiltrating his teams of audio officers into seemingly impossible targets, such as Soviet bloc embassies and foreign government offices, including the headquarters of hostile third world intelligence services, was his life's work. He had taken the Watergate debacle especially hard on a personal level: Although ex-Agency officers Howard Hunt and James McCord, and former FBI Special Agent G. Gordon Liddy of the White House Plumbers had never been members of Darth's elite team, he felt that their bungled Watergate penetration had brought shame on the CIA's audio operations.

Missions in the field demanded risk-taking, even in the midst of this politically charged atmosphere. An enterprising case officer in the Far East had recruited the landlord of an important Communist embassy in a South Asian country. The building was temporarily unoccupied because the embassy staff had been evacuated during one of the region's border wars, and the landlord was willing to "host" Schraider's audio team, provided they could enter the compound with him in daylight. The *chokadar* night guard had not been recruited into the scheme, and the local cops kept a close watch for burglars after dark. The Communist diplomats were not scheduled back in the capital for several weeks, giving Darth's audio wizards ample time to seed the building (including the living quarters of all the staff) with listening devices.

He had succinctly outlined the problem, but was leaving the solution up to me. "Take your pick of these assholes," Darth said with a proud smile as he pointed his swagger stick toward a half dozen officers seated around the cavernous office on the overstuffed brown leather furniture, which could only be found in the executive suite of a supergrade officer.

His manner made me acutely aware of the fact that I was just a lowly disguise officer who could hardly be expected to drive a hard bargain with such an Agency icon. Yet when I'd heard of his pending operation, I had suggested my disguise team work with Darth's section on equal footing. It was my first foray into big-league Pinball; the opportunity had been there, and I'd seized it.

A quick scan of the room confirmed that I was not a member of their fraternity. I looked into the hard eyes of some of the best electronics experts and second-story men in the business. I recognized the Chief of the Surreptitious Entry Unit and his deputy, whose expertise involved swiftly defeating any lock or safe known to man. Among the other burly men, I identified several crack audio officers, renowned for operations in which they'd slipped over high walls or scaled icy rooftops in the Soviet bloc under cover of darkness to penetrate target buildings. They had sometimes remained inside for days, while they installed hidden microphones and transmitters designed to eavesdrop undetected for years.

Unknown to the public, the men in this room were the glamour boys of OTS. Their equipment was beyond state-of-the-art in terms of electronic miniaturization and power-source duration. They could plant a bug that hopped across a range of frequencies almost impossible to detect, yet consumed less power than a quartz watch. Their esprit de corps and autonomy were legendary. They earned promotions rapidly and routinely received commendations.

I opened my briefcase and removed an 8 × 10 photograph of myself in a new generation of the GAMBIT disguise that could easily transform any of them into whatever persona was required for this operation. I had used this disguise to quickly change into a bazaar wallah, right down to the kurta and the soiled white turban. "I wore this rig a couple of times in the field," I said, "tailing and photographing a KGB officer who was trying to pitch a code clerk from a friendly embassy. I think you'll agree I don't look like your typical American."

The hardened audio officers seemed impressed.

"Imagine that you're wearing one of these in the cab of a repair van that the landlord waves through the front gate of the embassy compound," I said, then turned to Darth Schraider. "Broad daylight presents no problem."

I sensed that the atmosphere of skepticism was gradually changing to one of credibility and rapt attention.

"We can procure and prepare all the materials you'll need, and, if you have the resources, we can use the best consultants," I told Schraider. "But I'll ultimately want my most experienced disguise officer on site with the team. And he has to be prepared to go into the target if necessary to maintain the integrity of the materials, especially in the monsoon with no air conditioning." I was imposing an unprecedented condition. Darth's people operated alone. But they all recognized that my recommendation provided the operation with added insurance. "We can get to work right away ordering custom materials and making the disguises and have at least three of your officers ready to travel by Friday." I smiled benevolently at the audio boys, then looked back at Darth with what I hoped was youthful innocence. "I assume, sir, that you have the budget to make all this happen."

Ten minutes later, I strolled into the office of my boss, Tim, on the second floor of the Central Building. "I just got another fifty thousand dollars for disguise development," I said, trying to keep a straight face. Tim heard my words, but their meaning did not immediately register. How could I make a simple verbal request, not supported by countless memos and endless meetings, and score fifty grand? In my head, I heard the big Pinball board clank and chime—I was learning to play the game.

Two weeks after the meeting in Schraider's office, his audio penetration team, accompanied by one of my best disguise officers and the local landlord, entered the Communist embassy compound early on a steamy monsoon Sunday morning. They came out three days later. The devices they planted would continue to transmit valuable intelligence for years to come.

Headquarters, April 1976 ■ By the mid-1970s, the serio-comic confrontations between Bull and the technical gurus at Headquarters had reached epic dimensions. But the dynamics of conflict between the case officers in the stressful world of field HUMINT and the technical experts in the sterile laboratory or workshop would soon give way to a system of demand and supply that improved the performance of both sides. Through bold looting forays back in Washington that Bull and his like-minded, workhorse chiefs of station conducted, a more flexible, cooperative relationship between OTS and the field eventually developed.

The new watchwords were no longer "operations first" or "can do." Instead, the key question was, "Is this operation worth the expense in both money and technical expertise?" OTS officers were trained to ask: "What is the operational goal?"

When challenged to justify the need for his often outrageous technical demands, for example, Bull would produce elegantly formal request memoranda, with all the proper operational and bureaucratic bows neatly tied. These requests evolved into a system to justify not only the technical support but the field operation itself within the larger scheme of Agency programs. Ironically, it was a traditional Agency case officer like Bull Monahan who would help transform high-stakes Pinball into a mini-revolution in the Clandestine Service, opening the path for more programmatic thinking. The "old boy" system was replaced by a more level playing field, and we began to plan ahead.

This shift was fabulous for my business. Bull Monahan, a near-genius brimming with ideas, loved the world of disguise. We reached a tacit understanding that he would keep applying the pressure on OTS for improved disguise technology. In turn, I would discreetly keep him informed of any progress we made on promising new technologies that he might include in his request for field support.

To set this process in motion, I arranged a meeting between Bull, the ultimate user of the most innovative disguises, and Jerome Calloway, the ultimate source.

Burbank, California, May 1976 ■ Bull was straining to remain still in the makeup chair, while I did my best to finish applying the delicate FINESSE material on his left cheek. Always impatient, Bull insisted on talking to Jerome, who was poised nearby, seated backward astride a tall lab chair. We had been working for almost an hour in Jerome's private makeup studio in the three-car garage of his home.

Jerome leaned forward, his hairy forearms across the back of the chair, holding a smoldering Salem between the tar-stained thumb and forefinger of his left hand. The brilliant southern California spring sunshine flooded Jerome's lush garden outside the open garage doors.

"This is the material Jerome began using a while back for these subtle adaptations," I explained to Bull. "He'd forgotten about it till I told him the idea in your last tasking cable. Then he found the stuff in a cigar box up on that shelf."

Bull's restless mind had hatched yet another audacious concept: We did not realize it at the time, but he was upping the ante on the FINESSE materials and would help lead us to another, more effective breakthrough, which would later be code-named DAGGER.

Jerome took the cigar box from his workbench and held up some of the makeup material. "I came up with this working on *Kid Gallahad*," Jerome said. "John Garfield was always trying to push the visual envelope of his character so we had to invent things on the fly. Garfield wanted a close-up of blood spurting out of his broken nose when the stunt man hit him. So his makeup guy came over here between takes and we threw together this particular formula. Worked like a charm."

We all laughed. Garfield's portrayal of a battered prize fighter was considered revolutionary for its gritty visual realism.

"In fact," Jerome continued, gesturing at the racks of shelves stretching from floor to ceiling, holding hundreds of lab jars with makeup devices that looked like the body parts of monsters, "most of this stuff was invented that way. The actor and director need something, I do my best to provide it."

Jerome puttered around his storage shelves, selecting some choice examples of humanoids and various facial appliances that had distorted the well-known faces of Hollywood's greatest icons for several decades. "The other makeup men in town come over and buy the extra pieces I cast from my molds," he said. "This workshop is the motion picture industry version of a boutique."

Bull chortled, his heavy jowls shaking. Luckily, I managed not to smear his makeup with my adhesive brush. "Yeah, we have our own version of a boutique in the spy business, too, don't we, Mendez?"

Bull was in a fine mood. He and Jerome had hit it off immediately. Both men were second-generation Irish-Americans, proud of their heritage. To my surprise, Bull had revealed that he had done his undergraduate degree in theater arts and was a "frustrated actor," probably an ideal education for a spy. His daughter would graduate that June with a degree in theater, and Jerome had graciously offered to arrange some auditions.

I gave Bull a hand mirror and pulled off the barber's cape so he could stand up and examine himself in the three-sided tailor's fitting mirror near the lab bench. Bull leaned forward, then did a slow pirouette, carefully checking his profile from all angles. "This looks pretty good, but I want the cheekbones and the chin much more prominent to distract people from other things I might be doing." I had already transformed Bull into a near-gargoyle, but he wanted an even more outrageous appearance. As always, he had a firm grasp of the larger operational issues involved: In Bull's part of Western Europe, it was considered the ultimate faux pas to stare at people with unpleasant features or physical disabilities.

Knowing Bull, however, I was afraid he would want to push this social reality to the extreme. "I'll try," I stalled, "but we don't want to get ridiculous, either. Let's see how long you can wear it comfortably out at lunch. Jerome and I will watch how the light works on this new material."

It was a short drive from Jerome's home to the Universal Studios cafeteria. I held back a grin as I watched Bull, who seemed more nervous wearing his new face among these celebrities than he would have been shaking surveillance in China or Berlin. We were indeed subjected to some close scrutiny from "big names"; as an award-winning member of the industry, Jerome had to pass a gamut of shmoozers on the way to our table.

When we climbed back into Jerome's gleaming yellow Pontiac Brougham after lunch, Bull seemed thoughtful. In effect, he'd just passed an operational test, eating a Caesar salad and tuna filet in full disguise in front of the world's experts on illusion. "You might suggest to your chemist that they put a little more elasticiser in their formula, Tony," he said, shifting his strained jaw from side to side. "If we're going to make the chin and cheekbones more prominent, the whole rig's going to have to be a lot easier to wear for extended periods."

At Jerome's garage lab, Bull opened boxes and jars, fingering the strange collection of colorful materials as he hummed a tune off-pitch. "When I come back this summer, I'd like you to be prepared with a whole new disguise . . . glasses, teeth, the works. And I want to learn how to actually make those new pieces and apply them myself."

That was classic Bull Monahan, always making grandiose requests. But I realized he would be one of my strongest allies, especially if his energetic disguise initiatives helped pull off an ambitious operation he was not authorized to discuss, due to the need-to-know principle.

Bull wandered restlessly through the lab. "And, Jerome, maybe you could get me one of those cobweb machines you mentioned this morning." He rubbed his hands in delight. "That's just the gadget we've been looking for to cover our entry into an audio target through a wine cellar door that hasn't been opened in fifty years."

Jerome shared Bull's pleasure at the prospect of American spies using another movie illusion. "No problem, Bull. The special effects guys sell one in a nice little carrying case. It's basically just a quarter-inch electric drill with a fan blade hooked up to a dispenser of special glue. You can cover a whole room with cobwebs in a couple of minutes. Let me know if you need anything else," Jerome added. "Maybe you'll want to make a snowstorm, or part the Red Sea."

"You never can tell," Bull said with a wicked grin.

8 ■ Moscow Rules

. . . in the belly of the beast . . .
—"Jacob Jordan"

Moscow, February 1976 ■ Jacob and I crunched across the rutted snow through swirls of bus exhaust, following the herd of Air France passengers into the Sheremetyevo Airport terminal and down the stairs to passport control. Outside, it was a bright, frigid Russian morning. In the immigration lines, the atmosphere was stifling, the lighting dim, and the atmosphere tense. Returning Soviet citizens were probably wondering how they would talk their way through Customs with their carry-on bags stuffed with whiskey and cigarettes from the duty-free counters of Paris.

I had my own share of anxieties. Although I'd conducted dozens of operations using alias documents, this was my very first trip into the

Soviet Union, and I was worried that the alias legend in my burgundy official passport might not be convincing enough.

The previous summer, I had been sitting in my Foggy Bottom office when an officer on the Agency's Counterintelligence Staff called.

"Bad news, Tony," he'd said. "North Vietnamese intelligence has your true name and technical specialty, and they've passed it on to the KGB."

I felt the proverbial chill race down my back and let out an angry, frustrated sigh. When Saigon had fallen that April, someone with whom I'd worked had obviously been left behind; under interrogation, that person had revealed his knowledge of the CIA. Now my name, description, and professional capabilities were on the list of "enemies of the state" on file at the KGB's Moscow Center on Dzerzhinsky Square and in every *residentura* in Soviet diplomatic missions worldwide. The risk of being identified immediately by skillful Soviet counterintelligence officers, especially when applying for a visa or trying to operate here in Moscow, was high. Such knowledge was certain to give heartburn to any operations officer, and I was no exception. My alias and the minimal disguise I wore certainly wouldn't conceal the fact that I was CIA, but hopefully it would keep the KGB in the dark about my true identity and specialty.

The lines shuffled toward the smeared Plexiglas cubicles, where young uniformed KGB Border Guards carefully inspected each passenger's documents with a slow, detached interest. I knew their apparent apathy concealed intense professional scrutiny. Just before stamping an entry cachet, I noticed, each Border Guard stared directly into the eyes of the traveler, the usual ploy meant to detect undue anxiety at an especially vulnerable moment. Between every two booths, an older KGB officer, wearing the standard dark baggy suit of his caste, stood like a mute sentinel, his eyes sweeping the passengers. Occasionally, a plain-clothes man would nod, and his uniformed colleague would politely "invite" someone into a screened examination area to the right of the echoing hall.

I watched a rotund Russian in a horse-blanket overcoat and well-worn fox shapka nervously shepherd his henna-haired wife into a cubicle. The man had enough clout to travel to the West with his wife, unescorted and on a foreign carrier, yet he was not a high-ranking member of the *nomenklatura*, who would have been quickly ushered through the VIP formalities in a separate terminal. So who was he? If this Russian was actually a KGB officer working under TASS or trade mission cover, his legend would have required he run this gauntlet like any other returning citizen. Good tradecraft required that he live that legend fully, even here in Moscow, in the event that Western intelligence had been resourceful enough to keep surveillance on him aboard the plane from Paris. Was the woman really his wife? I wondered. Maybe she was the intelligence officer, and he was simply window dressing.

Welcome to Moscow, I mused, where appearance often masked sinister reality, a city that defined the well-known wilderness of mirrors.

I would never find out if the man was an authentic worker bee of the vast Soviet bureaucracy or a highly competent spy. At that moment, Jacob passed through immigration control ahead of me, and then I entered the booth. The young inspector's uniform was well pressed, his collar and fingernails clean. He took his time flipping through my passport, pausing to study the multicolored foldout "service" visa issued by the Soviet embassy in Washington with my photo attached. Then he examined my airline tickets, which had come from the State Department's travel office. I was sweating profusely, as much from the steamy heat as I was from the pressure of the situation. This was exactly where our expertise was tested and enhanced—the chokepoint of airport security controls. We could absorb any strength or weakness in the systems that we had spent our entire careers probing and challenging.

"What is purpose of visit?" he recited, using a standard phrase he might not have even understood.

Was there a hidden microphone underneath the counter to record a telltale voice print of "suspicious" characters like myself? That notion did not stem from simple paranoia. The KGB's Seventh Chief Directorate, responsible for surveillance throughout the Soviet Union, managed an incredibly elaborate and innovative eavesdropping apparatus. This Directorate and its counterintelligence analog, the Second Chief Directorate, had virtually unlimited personnel and funding resources available to thwart foreign intelligence efforts, especially here in Moscow, the "Center."

"Temporary duty at the American embassy, sir," I replied. Jacob and I were working under the cover of one of the minor alphabet soup agencies that had burgeoned in Washington since the 1950s. Our legends had us pinned as low-level administrative types, on par with bookkeepers or motor-pool inventory specialists. Adopting the protective coloration of such drones, we hoped, would free us from the intense surveillance endured by some American officials who were permanently assigned here under diplomatic cover.

The inspector removed part of the foldout visa and sat poised, with his rubber cachet stamp raised, then shot me a quick, searching stare. I tried my best not to flinch.

"Have good visit in Soviet Union," he said woodenly, hammering down the stamp.

THE EMBASSY VEHICLE was a salt-rimed Chevy station wagon that looked as if it had lost the war against Moscow's potholes and ice ruts. The driver, a Russian in his thirties with a middleweight's body turning soft, drove the car as if it were a tank, slamming on his horn and weaving through the lines of trucks and rust-pocked little Zhigulis, the Soviet equivalent of Fiat econo-boxes. He accepted a pack of Marlboro 100s with a grunt and chain-smoked all the way into the city along Lenin-gradskoye Highway. Occasionally, he'd point a finger at some fleeting, snow-covered point of interest. ". . . Khimki . . . Ring Road . . ."

Our low status as Temporary Duty (TDY) nondiplomats had obviously not sparked his interest. This suited us fine. The man belonged to the embassy's Miscellaneous Services Section, officially a branch of the Soviet Foreign Ministry's service arm, the UPDK. In reality, the UPDK was controlled by the KGB's surveillance and counterintelligence directorates. If he had been suspicious, he probably would have drawn us into conversation, no doubt revealing in the process an unusual fluency in English.

We arrived at the American embassy on Tchaikovsky Street along the Garden Ring just before the lunch hour. My initial reaction to the 1950s-vintage, former apartment block was one of foreboding. Maybe it was my critical painter's eye, but I thought the place looked grimly misshapen, and even the ocher stucco facade seemed poorly constructed, discolored in places with mismatched patches. The station wagon lurched over an icy hump of unswept sand choking the entrance, then past a guard booth with a stern "mili-man," a member of the Interior Ministry's Militia civil police. We passed under an archway and emerged into a long, narrow courtyard.

When we entered the warren of the chancery building, I saw that my negative first impression had been justified. Even though the State Department had worked hard to transform the original apartment house into a working embassy, people assigned here still had to contend with cramped, low-ceilinged rooms, narrow, musty halls, and dim staircases that resembled rickety ships' ladders. The floors sagged and rippled in places because the planks had been laid on freshly felled tree trunks—one of the few commodities in abundant supply in the Gulag-haunted Soviet capital of the 1950s. The electricity was primitive, and when American specialists had been imported to renovate the offices, they were disgusted to find the insulation between walls was a gritty mixture of coal ash and sawdust, making the entire compound a potentially deadly firetrap.

But the Tchaikovsky Street embassy was an ideal site for electronic eavesdropping. Embassy and Agency security officers estimated that the KGB's ubiquitous local employees had seeded the entire building with hard-wire and wireless bugs. The windows were silently scanned with microwaves that could reproduce the vibration of conversations into usable recordings at the numerous KGB listening posts ringing the embassy. Between the overtly inquisitive UPDK local employees and the hidden bugs, Americans, from the ambassador to the lowest-ranking Marine security guard, were subject to audio surveillance during every moment they spent in the U.S. Mission, including in their apartments.

But there were important exceptions.

Because American officials in many embassies needed a secure area to discuss sensitive cases, they usually went to "the Bubble"—the generic term for a clear plastic-walled enclosure, raised from the floor on transparent Plexiglas blocks and meticulously cleaned only by American hands, so that none of the local staff, no matter how ingenious they were, could attach miniature listening devices to the structure without being detected.

Eventually, Jacob and I were expected at a meeting with the local CIA chief, "Bill Fuller," a man with whom we had served in the Far East and whom we knew as a very imaginative and progressive case officer. His deputy, "Jacques Dumas," a feisty Marine Corps veteran, with a Harvard degree and excellent Chinese and Russian language skills, would also be with him. But before we could attend that meeting, Jacob and I had to display our deceptive cover through the dim corridors of the chancery, where the KGB snoops could have a good look at us. UPDK had assigned us a shared room in the Peking Hotel, located up the Garden Ring from the embassy, a rather spartan establishment befitting our low rank.

That afternoon, after a greasy daily special lunch at the embassy cafeteria, Jacob and I unpacked our briefcases bulging with authentic, unclassified paperwork, and settled down to our separate drudge jobs. As the tedious afternoon passed, several Russian local employees found excuses to visit the two offices assigned to us. As I carefully reviewed monotonous files through a pair of "plano"-lensed (uncorrected) horn-rimmed spectacles, I sensed with satisfaction that my change of appearance was subtle but effective, and that we'd passed initial muster. I was also wickedly amused that my debonair former mentor, Jacob, had been obliged to surrender his tailored Bond Street wardrobe for Sears's finest polyester. I had myself opted for striped, pointed-collar dress shirts and flamboyant, wide ties at least a couple years out of fashion to match my older, seedy appearance, complete with graying temples and a dental appliance that included a gold-trimmed incisor.

Scanning the meaningless columns of budget figures, I thought about the circumstances that had brought the two of us here. I'd been hard at work as Chief of Disguise at Headquarters a few weeks earlier, pushing to expand my program, when a cable had come from Moscow requesting OTS experts to conduct a thorough disguise survey in the Soviet capital. Jacob had been back in the field for more than a year, leading a contingent of technical operations officers responsible for disguise and alias documentation support in the Soviet bloc. Based in the West, he'd been up to his old tricks, facilitating some of the more daring infiltrations and exfiltrations, this time through the Iron Curtain. We'd both shot cables back to Moscow volunteering our services. After a flurry of subsequent cables from the field and Headquarters, it was agreed that we would go as a team.

The Soviet Union was the ultimate professional destination for any CIA operations officer. Even though we'd been friends and colleagues for many years, Jacob had not been thrilled when I'd appeared on the scene of this inaugural OTS disguise survey in Moscow. But when we met in Paris and spent an evening catching up, enjoying a memorable meal of goose, sausage, and cheese at a bistro on the Île St.-Louis, accompanied by several bottles of Cahors wine and snifters of Armagnac, we both agreed that it was "obvious" our operational skills made us an ideal match, and it was silly to engage in turf wars.

"We're actually going to Moscow, Tony," Jacob said, hoisting his brandy snifter in the air. "The belly of the beast."

Now, here we were, perched at our Bob Cratchit desks, when the last of the local employee snoops pulled on her galoshes and departed. After waiting another fifteen minutes, Jacob tore a single piece of paper from a legal pad, laid it on a glass-topped table, and scrawled a message: "Time to go to work."

We took a circuitous route to the Bubble, mumbling that we were searching for the back stairs to the rear courtyard community center, site of the evening happy hour. Jacques met us at the outer door and punched in the keypad lock code. Only when we were all inside the chilly, air-conditioned plastic box could we speak freely.

"Welcome to Moscow," Fuller exclaimed, "Wimbledon Center Court."

Jacob grinned at our surroundings. "The 'Cone of Silence,'" he quipped, referring to the fanciful antisurveillance technology in the *Get Smart* TV series.

"We use it here all the time," Jacques said, "especially for meetings like this. Otherwise, we'd have to pass notes at the water fountain."

Bill gave us a one-page, EYES ONLY memo outlining the CIA's Moscow rules of engagement with the KGB. "For the moment," he said, "this is our Bible. We'll ask you to follow these rules to the letter when you're out there on the street."

The Moscow rules, Bill explained, had evolved over almost twenty-five years, subsequent to the Stalinist deep freeze, when the Soviet dictator had decreed that Western embassies move several kilometers from the Kremlin. Most had clustered here in prerevolution villas, on quiet streets leading like spokes from the Kremlin to the wheel of the Garden Ring. This concentration of target embassies, Fuller added, had allowed the KGB's Seventh Chief Directorate to marshal their surveillance forces in a relatively small sector of the sprawling Soviet capital, just as Washington's "Embassy Row" along the 16th Street–Massachusetts Avenue corridors provided a similar advantage to the FBI. To increase the effectiveness of their surveillance net, the Soviets imposed rigid travel restrictions on foreign diplomats, journalists, and business people. Americans could not travel outside Moscow or Leningrad without written permission. The State Department placed reciprocal constraints on the Soviets working in America, but KGB officers working under United Nations, trade mission, or TASS cover managed to move around the United States with relative freedom.

"Rule One," Jacques said, tapping the memo. "Assume *every* Soviet you encounter is connected to a larger surveillance apparatus. This includes the women shoveling snow in the winter and the guy selling ice cream in Gorky Park. The ticket-taker at the zoo reports to the KGB. The bartenders in every hard-currency bar and restaurant are on the payroll of the Seventh Chief Directorate. Half the taxis in this part of the city are driven by their men."

"In short," Bill added, "we assume constant surveillance."

This saturation level of surveillance, which far surpassed anything Western intelligence services attempted in their own democratic societies, had greatly constrained CIA operations in Moscow for decades. The KGB always made a concerted effort to identify any Agency case officer assigned to Moscow, then kept him and his family under tight surveillance throughout their entire tour. This extreme level of hostile scrutiny was unique, even for the Soviet bloc. A skilled officer operating in Prague, Budapest, or Warsaw, for example, might use clever tradecraft to slip his surveillance periodically for secure meetings with important assets. But conditions were so tight in Moscow, Bill and Jacques explained, that such meetings were rare; an officer might never hold one during his entire two-year tour.

However, Agency case officers in Moscow were still expected to perform the most important duty of clandestine espionage operations: secure and timely communication with agents-in-place. The two station officers reviewed for us the clandestine agent communications plans they had devised to keep channels open to their assets in such a hostile area.

"We have to spend every waking moment working on these problems," Jacques said. "Everything we do outside our apartments and these offices is geared toward agent communication."

"I haven't taken an unplanned stroll on the street or had a friendly tennis match for almost two years," Bill conceded. Whenever he was outside the "sanctuary" of the American mission, he was constantly at work, trying to overcome inevitable threats to our agent pipelines by playing mind games with the far superior forces of the KGB.

Turning back to the memo, Jacques described the basic structure and reputed modus operandi of the Seventh Chief Directorate's street surveillance teams. Once assigned to a suspected foreign intelligence officer, a dedicated team focused their entire attention on that person, twenty-four hours a day. Identifying senior CIA officers was sometimes made easier by the Agency and State Department's practice of giving them fairly senior diplomatic cover jobs so that they had plausible reasons to visit a variety of Soviet government offices, meeting and assessing potential Soviet official targets on the diplomatic social circuit as well. Devoting around-the-clock surveillance to an officer who was merely suspected of espionage was not simply an extravagance for the KGB: They understood the immeasurable harm to the Soviet Union that an effective spy could inflict.

When Jacob and I reached Moscow in 1976, an uneasy equilibrium existed in the spy-versus-spy power struggle. The KGB could not be certain of how many foreign intelligence officers were working in the capital, nor were they sure of exactly who they all were. To protect themselves, the Seventh Chief Directorate tended to overestimate the numbers of the opposition, and to saturate people they considered obvious candidates with grossly inflated surveillance teams. The dubious distinction of being targeted for concentrated surveillance arose from several factors. It was important for us to identify what they were in order to avoid coming under suspicion as we conducted our disguise survey.

"We know from our reporting and defectors that they study our overt behavior patterns, overall demeanor, and daily profiles," Jacques explained. "So you'll have to be very careful about your actions in this regard."

Bill added that the sheer size of the KGB surveillance operation often made it cumbersome and less flexible than it had been perhaps fifteen years earlier. This deterioration was due to several factors. By 1976, both the foreign diplomatic presence and business community had grown considerably since the darkest days of the Cold War. Our closest NATO allies, Great Britain and West Germany, also had large embassies in Moscow and were subject to the same level of scrutiny and suspicion. Then there were the Chinese, hardly the stalwart allies of the Soviet Union they had once been. Finally, Moscow's population had steadily increased, year by year, despite official Soviet efforts to restrain migration to the coveted Center.

To meet these challenges, the KGB had played its hand like a gutsy table-stakes poker player, raising each of the opposition's bets. As the CIA and the West German Bundesnachrichtendienst (BND) intelligence service expanded their Moscow operations, for example, KGB surveillance teams proliferated. The Soviets were good but not "ten feet tall." The increasing density of surveillance trailing foreign suspects from the Western diplomatic district west of the Kremlin occasionally led to one team tripping over another in pursuit of their quarry.

"That can be damned funny," Bill said. "But it doesn't happen very often. These guys are generally invisible."

"And that's the problem," Jacques admitted. "Your average Soviet citizen can somehow sense the KGB in the Metro or the queue for the trolley, but they just look away. You grow up in this country, you acquire a set of antennae to detect KGB vibrations two blocks away. I wish we could do it that easily."

"Let's look at how we think the bad guys operate," Bill suggested, giving us another single sheet showing a schematic diagram. "This is the estimated size and MO of a typical KGB surveillance team." Inside the circle at the center of the page was "the rabbit," a suspected CIA officer. Depending on his cover—diplomatic, lower embassy rank, or nongovernmental—the officer might routinely have a sizable team of surveillance specialists serving in shifts around the clock, with a rotating stable of cars at their disposal. The team's function was to keep the suspected officer in direct sight, or to be confident that the rabbit was safely ensconced in his office or otherwise accounted for at all other times.

"By the way," Bill said with a wry grin, "we're pretty sure our Soviet friends have augmented their audio bugs with hidden audio and video in the apartment blocks, probably trying to acquire some spicy 'Peyton Place' tape for potential blackmail."

"Not only do the walls have ears," Jacques added with a chuckle, "they also have beady, bloodshot eyes."

For a moment, I considered the staggering logistics load involved in such a surveillance operation. Beyond the large contingent of full-time members of the team dedicated to a single suspect, you also had all the language-qualified listening-post monitors, along with the analysts, who examined all this information.

At the pavement level, the multiple surveillance teams, each targeted on their individual rabbit, had to stay hidden nearby—especially important in the near-arctic Moscow winter, when nocturnal Fahrenheit temperatures often fell to thirty below zero.

"They've got to have 'warming rooms' the size of a large police barracks in buildings up and down the street," Bill explained. In fact, we suspected the closest one was in the apartment block abutting the chancery building to the south.

When the night shift was reasonably confident that their target rabbit was either still working or had turned in, they took their own breaks, but someone always maintained observation on the arched entrances leading into the embassy courtyard or the diplomatic ghetto, a complex of high-rise apartments on Kutuzovsky Prospekt.

Beyond the surveillance teams dedicated to an individual officer, there could be mobile supplementals to enhance coverage if a suspect seemed to be going "operational." These additional teams could be marshaled by radio, especially if a suspected intelligence officer suddenly made a "provocative" move, such as running a red light or bolting on foot across lanes of traffic to dash into a Metro tunnel. This type of action was absolutely against Moscow rules, unless it became necessary for officers near the end of their tours to try to break surveillance this way.

"We're up against a helluva opposition here," Bill admitted. His ranks were thin and had to hide in the greater numbers of the American community. In order not to tip their hand, CIA officers tried to maintain a seemingly innocuous cover-job pattern and demeanor, sticking to it over months and years. But they also had to be flexible enough to meet a variety of operational problems. For example, a hypothetical officer whose official cover job involved trade and agricultural commodities may have to spend soul-deadening months in the drab, overheated offices of his Soviet counterparts, driving to and from these meetings each day and seldom deviating from his route. His social life and recreation also had to fall into established patterns, perhaps consisting of weekly bridge games at another embassy and cross-country skiing on winter weekends.

Once this officer was reasonably confident that he had dulled the edge of potential surveillance, he might engage in several unalarming operational activities, such as servicing dead drops on a single, well-planned cover outing, Jacques noted. "But these have to be carefully considered and rehearsed," he warned.

A good officer could perform some of these tasks in full view of surveillance, but others required a fleeting moment of privacy. Monahan had taught me years earlier that a spy could be out of sight momentarily but still not alarm the surveillants tailing him. There was a very subtle and complex psychology at work in such deception. The operational act might be as simple as dropping an empty cigarette package into a trash can in the spotless Belorusskaya Metro station, or chalking a tiny cyrillic "D" in a phone kiosk along Gorky Street. Performing these bland acts was not a major challenge, but accomplishing them in a manner that would not raise suspicion through an obvious shift in demeanor required a level of tradecraft only the best case officers possessed.

However, it was one thing to practice sound countersurveillance techniques in the face of the clumsy Gendarmerie in Dakar, or even their more sophisticated counterparts in Damascus, and quite another to attempt to escape from the relentless coverage of the Seventh Chief Directorate on the streets of Moscow. As the four of us in this plastic cubicle understood all too clearly, the stakes were immeasurably higher here. If a case officer's behavior on the street was obviously "provocative" and included blatant surveillance detection runs, quick detours into multiexit establishments like the GUM department store, or sudden backtracking on the Metro, the KGB would conclude that they'd just confirmed the officer's CIA identity and would intensify surveillance to identify any Soviet citizen he might contact.

This grim reality lay at the core of the problem facing our Moscow operation. Unable to conduct clandestine personal meetings with our important agents-in-place, we had instead to rely on "impersonal communications" contact, which could be skillfully manipulated by the KGB and packed with misinformation if our putative agent-in-place had long since been rolled up.

"We just have to find some way to securely conduct personal meetings right here in Moscow," Jacques said, summarizing the nub of the problem. "That's where your OTS disguise technique might play a role."

I was afraid that Bill and his deputy were expecting a single, magical solution, whereas I knew from experience that physical disguise was only one element in a complex tradecraft problem. "We'll be happy to conduct a good disguise survey," I told them. "But to do so, we're going to have to carefully analyze your entire operation, everyone's positions and profiles, and also dissect the opposition's techniques."

Bill and Jacques exchanged glances. After all, Jacob and I were from OTS, and we were not graduates of the Internal Operations course, the DO's elite training for assignments like this. But both Moscow officers understood we had unique expertise and experiences to offer—tradecraft they desperately needed to meet the challenge.

"Well," Jacques said with a grin that broke the tension. "If you weren't making this sacrifice to save freedom, democracy, and motherhood, you'd probably be out robbing banks. Have at it."

It was old Vietnam gallows humor and we all laughed. It was good to know we had a solid crew in Moscow, because Jacob and I would be in and out of this grim city for months, helping to overcome one of the toughest operational problems the Clandestine Service had ever faced.

ON OUR LAST day in Moscow several weeks later, we celebrated by leaving our dingy cover office early and strolling along the Arbat among the street vendors, then all the way to Karl Marx Prospekt on the northwest corner of the Kremlin. The clear early afternoon light quickly faded into the violet dusk of winter, reminding us that Moscow lay on the same latitude as Alaska. By now, we were equipped with fox shapkas and thick mittens. Reaching the snowbound Alexander Gardens, we decided to take a final "tourism" stroll, ambling counterclockwise around the medieval fortress, down to the Kremlin embankment and along the high, two-kilometer russet perimeter walls back up Borovitsky Hill from the river. We headed toward the candy-striped onion domes and spires of St. Basil's Cathedral, now emerging from a sudden snow squall. As we entered Red Square, the enormous clock on the Spasskaya Tower inside the Kremlin walls announced that it was five P.M., and another horizontal band of snow almost swallowed the ruby star atop the spire. The honor guard came goose-stepping out of the archway and proceeded to Lenin's tomb in their hourly ritual.

Despite the harsh weather, there were lines of reverent visitors snaking toward the dark granite monolith of Lenin's mausoleum, pressed up against the Kremlin's outer walls and dominating the center of Red Square. But aside from the soldiers and the faithful throngs waiting in the snow to pay homage to Lenin's waxy cadaver, the vast square seemed deserted.

"They must be hanging back this afternoon," Jacob muttered from inside his muffler, now drawn up to his nose.

I hazarded a couple of clicks with my Spotmatic, confirming the blue Zhiguli we'd sighted twice on the walk had disappeared. That could mean almost anything. Perhaps our unusual early departure from the embassy had tripped a half-hearted response, with a small, low-priority team tailing us simply to confirm if we were indeed "persons of little interest." According to a KGB defector, that was the term his Seventh

Chief Directorate colleagues used to describe the clerks, secretaries, assorted bean-counters, and "admin types" responsible for the daily housekeeping of the U.S. Mission. If so, that was good news, indicating our weeks of living an exceptionally dull cover had paid off.

Crossing Karl Marx Prospekt through the pedestrian tunnel, I casually scanned the faces of people moving toward us from the Metro stations beyond. Instinctively, I was trying to determine if any of these people appeared familiar from earlier encounters. I knew Jacob was doing the same, although he managed to keep up a lively commentary on the performance of *Giselle* we had seen the night before at the Bolshoi Ballet. True to our established pattern, we entered the cozy mahogany-and-brass "dollar bar" on the second floor of the old National Hotel and hung our topcoats and shapkas at a nearby booth. Vladimir, the barman, was a jovial fellow who sported a borsht-flecked necktie, emblazoned with the Courvoisier logo, probably a gift from a French salesman—or perhaps it was a French case officer from the SDECE intelligence service working undercover as a booze-peddling salesman. Had I been infected by Moscow's paranoia? You bet. Paranoia became a part of you in a society like this. I grew accustomed to it, as if it were a second skin.

"Vodka juice!" Vladimir exclaimed, flashing his dazzling stainless steel teeth. Only uncultured American technicians would think of diluting 180-proof Siberian vodka with sour, canned Moroccan orange juice. But he accepted our five-dollar bill for the drink and a refill with hearty good humor. "Tonight, be-yoo-tiful Russian music," he announced, pointing toward the dining room where a balalaika trio had launched into a twanging set.

Jacob downed his first screwdriver and grabbed the second. "Tonight," he replied, handing Vlad our printed American embassy reservation card, "we're eating dinner at the Praga."

This restaurant was one of Moscow's finest, housed in the mansion of a prerevolutionary duke, and reputedly owned, or at least controlled, by the KGB's counterintelligence Directorate. While a Muscovite had to have *blat* ("pull") to land a reservation, embassy employees were encouraged to dine there.

"Very nice," Vlad commented, fluttering his fingers over the invitation card as if it were a hot ticket. "Celebration?"

"We're going home." I smiled. "Our work's all finished. Tonight we take the Red Arrow to Leningrad for a few days duty at the consulate, then it's back to the land of the big PX."

Vladimir chortled. "I know big PX from Marine Guards. All my friends. All bring Vladimir Winston cigarettes and Ronson lighters. When you come back?"

He was making a convincing show of being a minor black marketeer, which, of course, made for excellent cover. "Whenever they need us," Jacob answered as we grabbed our coats.

Leaving the National Hotel, we were confident some faceless clerk in the Lubyanka would type yet another entry into our dossiers that night, confirming what the UPDK snoops had already reported about our Praga restaurant dinner reservations and our scheduled trip to Leningrad.

JACOB AND I followed the haughty uniformed doorman of the Praga, who led us into the elegant restaurant's main foyer, replete with a glowing crystal chandelier, gilt-framed mirrors, and a curving white marble staircase. He escorted us up the gleaming steps to the office of the Administrator, then motioned for us to take chairs along the wall.

The Administrator, a woman in her forties with lacquered hair and tobacco-stained teeth, sat behind her huge desk, energetically scolding an offending employee. The man stared at the floor, visibly trembling, and I wondered what his crime had been. When the "Red Queen" finally finished with him, she allowed him to leave with his head still on his shoulders. Only then did she turn to us and briefly inspect the invitation card the doorman had left on her desk. She snatched up a heavy telephone that looked like an artillery radio transmitter from the Battle of Stalingrad, and spoke brusquely into the receiver. Seconds later, a maître d' stood at attention in front of her desk, and she issued curt instructions.

The maître d' then guided us along palatial corridors past large private dining rooms echoing with the voices of rowdy patrons. I caught glimpses of burly *nomenklatura*, wearing the inevitable ill-fitting Brezhnev Special suits and uniforms sprinkled with medals. There were also a number of attractive young women in evening dress dining with these old crocodiles, an overt sign of decadence I hadn't expected. In several rooms, raucous New Orleans hot jazz or 1930s swing combos added to the cacophony.

We mounted another set of marble stairs and emerged into a tasteful winter garden, where an attentive waiter led us to a ringside table on the dance floor. The band played a fair imitation of Woody Herman, much to the delight of the gyrating Russians, many dancing wildly with no one in particular but having the time of their lives.

"Would the gentlemen like drinks?" our waiter asked courteously in fluent English. I had a sinking feeling that we were receiving special treatment, that we had indeed been expected and were right now under active surveillance.

Playing the unsophisticated bumpkins, Jacob turned to the waiter with a helpless grin. "What do you suggest?"

Moments later, we were spreading Beluga caviar on thinly sliced, well-buttered black bread and sipping icy Stolichnaya vodka, poured from a crystal flask into thimble glasses. As the waiter removed our borsht bowls and laid out the platter of flaky cheese tort and chicken Kiev, I pondered the serious operational problems we had identified during this initial survey.

SOON AFTER OUR arrival, Jacques had emphasized the fundamental element of operational life in Moscow and told us never to forget it: "By their very nature, Russians are distrustful." He reminded us that there were centuries of autocratic Russian history preceding the six decades of Communist dictatorship. "Lenin and Dzherzhinsky didn't invent the Secret Police," Jacques said. The tsar had maintained a dreaded security service, the Okhrana. But the Bolsheviks had quickly eliminated its officers, rather than incorporate them into the new Cheka. Under the tsar, the Bolsheviks had survived only because their clandestine skills and early tradecraft enabled them to escape the Okhrana. They turned these skills to advantage when establishing their own security police and espionage service.

Russians had been practicing the art of surveillance for hundreds of years. Their imaginations were rich with possibilities, and if they were traditionally distrustful of each other, they were doubly so of foreigners. Add to that the reality of life in Brezhnev's Soviet Union, in which trafficking in contraband and bribery were rampant, and you had a huge pool of well-qualified potential candidates from which to draw recruits to the surveillance Directorates.

And our Soviet opposition knew how vulnerable they were to high-level espionage. In the early 1960s, the KGB had been badly burned by the Penkovsky case. Colonel Oleg Penkovsky, an experienced officer in the Soviet GRU military intelligence, was a "walk-in" agent-in-place, with an ax to grind against his superiors, who had thwarted his career due to the fact that his father had been a Tsarist army officer who fought against the Bolsheviks in the Russian civil war. On a warm August night in 1960, he accosted a startled American tourist near Red Square and passed him a message for the American embassy, stating Penkovsky had information of "exceptionally great interest" that he wished to deliver to the CIA. The tiny Agency operation in Moscow that existed at the time was unable to contact Penkovsky, so he approached the British, who completed the recruitment. Within months, Penkovsky was delivering some of the highest quality intelligence that either the CIA or British MI6 had ever received in the Soviet Union.

Dispatched to London on a Soviet "research committee," Penkovsky underwent vigorous CIA-MI6 debriefing and agent training. He then revealed that Nikita Krushchev's Politburo planned on installing intermediate-range nuclear missiles in Castro's Cuba. Almost as an afterthought, Penkovsky turned over reams of detailed information on advanced Soviet weaponry.

Back in Moscow, this flood of high-grade intelligence continued, creating a classic case officer–agent liaison problem. If either CIA or MI6 officers operating under diplomatic cover regularly maintained contact with Penkovsky through dead drops or brush passes, they would eventually expose him to disclosure, yet both governments refused to stop or even slow the flow of secret intelligence Penkovsky was prepared to deliver in exchange for seemingly minor trinkets, such as gold watches, vintage cognac, and stylish fountain pens—all status symbols in austere, post-Stalinist Moscow.

The Americans and British reached a compromise: Janet Anne Chisholm, the wife of an MI6 officer operating under diplomatic cover in Moscow, became Penkovsky's action officer. The young mother of three, herself familiar with security procedures, often pushing a stroller, had brief encounters with Penkovsky in parks and crowded public buildings, where they exchanged rolls of film for his Minox spy camera, and she received handwritten copies of secret documents. By August 1961, Penkovsky had delivered so much material that the CIA had established a separate analysis shop, employing twenty translators.

This bonanza abruptly ended on November 2, 1961, just over a year after Penkovsky's recruitment. A phone emitted a prearranged number of rings in a Moscow CIA office, but there was no one on the line. Such a "dead telephone" signal meant that Penkovsky was loading a dead drop with critical information, perhaps indicating impending war. A case officer was dispatched directly to the drop. The moment he retrieved the matchbox drop container, four burly street operatives of the KGB wrestled him to the ground and forcibly searched him, ignoring his lame protests of diplomatic immunity. The officer was expelled within days.

Penkovsky's cover had been blown. TASS announced "the traitor Penkovsky's" execution on May 17, 1963. Although the terse bulletin did not specify the means, the Agency had ominous reports that GRU spies were generally chained to an iron cradle and fed alive, feet first, into a crematorium at Khodinka Airport near the Dynamo soccer stadium. With Penkovsky's capture, the elaborate Janet Chisholm agent-handling apparatus, run jointly by the British and the Americans, had been rolled up. The hemorrhage of critical intelligence information had also galvanized the KGB into building the unprecedented surveillance juggernaut that Jacob and I had observed over the previous two weeks.

In the fourteen years since Penkovsky's downfall, the KGB had become absolute masters of the operational tradecraft of deceptive surveillance in Moscow. They employed both static and mobile surveillance, and could call on a reserve force of MVD militia and alert Party member volunteers spread throughout the city in parks and Metro stations generally favored by spies. We also knew the KGB sometimes resorted to facial and clothing disguise in order to alter their patterns and profiles as they moved along the streets.

But in general, as Jacob and I had learned from Jacques's files and two weeks of active street work, the Seventh Chief Directorate teams preferred to maintain a discreet distance from their targets. We realized that this tendency might present an opening for our own relatively sophisticated disguise techniques, if their use was carefully planned.

After all, the KGB teams assigned to a suspected case officer might *seem* to have disappeared in the bustling traffic of the city streets, but they were never far away, and the sudden appearance of a strange "face" could be just as alarming as the disappearance of a target rabbit.

During relatively warm weather ten days into our stay, Jacob and I had spent the weekend exploring firsthand the surveillance techniques of the teams tailing Jacques and one of his colleagues, who were attending social events at nearby Western embassies. We did so in order to firmly establish the "macro" operational template—bureaucratese for the opposition's MO—on which any disguise program would either fail or succeed. Theirs was elaborate. The surveillance teams combined both mobile units and "foots," who rode as passengers in the cars and would bail out to move as pedestrians when required.

We had no trouble identifying the actual KGB cars, even though they did a good job of rotating from an impressively large pool. In Moscow at this time, most cars did not drive around with windshield wipers, which were hard to come by, expensive if available, and subject to theft. Most Russian car owners kept their wipers locked in the glove compartment and only snapped them on in the heaviest rain or snow. It was common to observe cars stopped at a red light, with their drivers hanging out from the doors to wipe slush from their windshields with rags. In a steady rain or snow, however, drivers pulled over and installed their precious wipers. A surveillance car, on the other hand, had to keep the rabbit in view. It could not afford to stop.

Jacob and I saw telltale wipers on a Zhiguli and a Volga rounding the corner of Karmanits Pereulok, a full block behind Jacques's sedan as he stopped to attend a cocktail party at the Philippine embassy.

"Observe the tires," Jacob said in his driest English manner. He pointed in the opposite direction as we ostensibly searched for the New Zealand embassy several blocks away. Both cars had new tires with well-chiseled treads, and each carried a driver and three passengers. Two of the men wore overcoats, shapkas, and ties, while the other two wore nylon parkas and wool ski caps—an unlikely mix in class-conscious Moscow.

Following form, the two vehicles circled the block, then split. When they returned, each had unloaded its parka-clad foots, who had taken up stations at opposite ends of the narrow side street that Jacques would have to use himself if he tried to move on foot to break surveillance. Although we had not seen any communications devices as obvious as "brick" radios, we did notice one of the foots talking into the lapel of his parka.

At our debriefing session that night, Jacob and I described to Jacques and his staff what we had observed, and the Moscow case officers responded with their own experiences. To our chagrin, these observations confirmed the validity of the current Moscow rules.

"Sometimes it seems they're working just over the horizon," a veteran female case officer noted. "It's like they're moving along parallel lines to your route. You can pick them out occasionally, but not very often."

"But you always know they're there," Jacques added. "They rotate around you as you travel, like a bicycle chain."

Given this hovering presence, the rules of engagement had evolved to condemn even seemingly innocent deviations from normal demeanor such as stopping to tie a shoelace or pausing to check a reflection in a store window. These actions were considered "peeking" and were forbidden to all case officers, as were more blatant acts such as jumping traffic lights, jaywalking, or driving against traffic on one-way streets.

"What happens if you try to buck surveillance?" I asked.

"They're on you like flies on cow manure," a gruff younger officer said.

"We're pretty sure a surveillance team gets punished if they lose track of one of us for any significant time," Jacques said. "If an American target just slips from sight for a couple of minutes, there's no problem. But if a target actually disappears out there on the streets, there's hell to pay. The entire surveillance group from the commanding officer down probably gets its pay docked and risks losing this cushy Moscow assignment unless they pull their socks up."

"The few times I've broken free," Bill Fuller commented, "I've had to pay for it. They were practically stepping on my heels for weeks afterward."

"It's essential that case officers allow the surveillance teams to maintain their 'zone of comfort,'" I agreed. "They have to preserve the cozy feeling that they know the target's whereabouts at all times, but remain unobserved themselves. If that comfort level is eroded, they'll close in so tight that even the simplest impersonal communication procedures will be impossible."

"But while we're keeping them comfortable," Jacques quipped bitterly, "they're effectively shutting us out of the action."

For the next few minutes, we carefully reviewed the clandestine impersonal commo procedures the Moscow officers employed. We needed to determine if these approaches could be expanded without breaking Moscow rules.

Using the telephone system for clandestine contacts was always risky in Moscow. Although there were automatic switching exchanges, all lines into the offices and apartments of foreigners were tapped and monitored around the clock by a virtual army of eavesdroppers. The "private" telephones of any Soviet citizen in a sensitive position were also tapped.

One-way radio was a much safer method of delivering instructions to a Soviet agent-in-place. Spy services had used this technique since the earliest days of long-range wireless broadcast. After World War II, many agents in the Soviet bloc were either recruited in the West, where they were trained and equipped, or by bridge agents among their own countrymen. Many Soviet assets had access to the one-time code pads needed to decipher encrypted one-way messages. Contrary to public perceptions of spies hunched with earphones in garrets deciphering messages, in real time one-way radio contacts could involve messages which the agent could tape on a commercial recorder and "break" when he felt absolutely secure. By the 1970s, when the Soviets were no longer regularly jamming foreign short-wave broadcasts, using these frequencies for one-way signals was considered both clandestine and secure, since the agent was the passive recipient of information.

Within the Soviet Union, radio broadcasts from, rather than to, an agent-in-place had fallen out of favor by the 1970s, and had, in fact, never been considered truly clandestine. Any spurious radio signal, no matter how brief, could be monitored by counterintelligence radio direction-finding units. If the agent continued sending radio messages, his transmitter could be pinpointed by DF triangulation. To thwart detection, agent transmissions had to be extremely brief and short-range, originating from continually shifting locations. But all agent radio communications entailed heightened risk because they required transmitters, code pads, site diagrams, and other compromising spy gear.

Secret writing and microdots were also very secure agent commo techniques with long pedigrees. The use of invisible ink, starting with goat's milk on parchment and lemon juice on paper dated back hundreds of years. By the 1970s, all major intelligence services had developed cameras that could photograph a full page of text and reduce it to the size of the dot on a fine-type letter "i," using negative techniques so that the dot itself appeared white against a white paper background. Microdots could be glued to the flap of an envelope or pasted inside the edge of a postcard. Although this form of communication was extremely secure, being caught with a microdot camera or other secret writing equipment was basically an admission of espionage, a capital crime in the Soviet Union.

All intelligence services used both dead drops and hand and vehicle "tosses" to pass physical objects without making direct contact between case officer and agent. In a "timed drop," the material stayed down only for a designated period before it was retrieved in order to prevent inadvertent discovery. Although seemingly secure, dead drops actually threatened extreme peril if hostile surveillance was both subtle and vigilant. If the drop or toss was detected and the site staked out, the agent would almost certainly be compromised, rolled up, and interrogated.

To be effective, dead drops also required complex and changing sets of signals. To alert an agent that "the drop is loaded," an officer might make a simple chalk mark on a light post along the route normally traveled by the agent. If the post was chosen well, the officer could place his signal while in full view of surveillance. Such signals themselves could be a form of impersonal communication, perhaps an answer to a certain question delivered by one-way shortwave radio.

In Moscow as elsewhere, it was part of the case officer's daily work load and discipline to be constantly on the alert for good drop and signal sites because of the need to change sites frequently and to rotate the inventory. Overuse of sites could establish a suspicious pattern. Case officers spent a lot of time on the street, not only locating likely sites but carefully surveying them for angles of approach and visibility from different perspectives.

Once signal and drop sites were chosen, a good deal of thought went into the type of containers to use in making drops and tosses. It was vital that they be fashioned to blend into the environment and survive the elements. But we simply could not use the beer can (or its local equivalent), which was ubiquitous in the West—in most parts of the world, scrap aluminum was scavenged almost as soon as it was dropped. In any case, there was not much solid trash on the streets of Moscow at that time because beer and soft drinks were sold in returnable bottles or straight from a tap.

OTS design engineers and craftsmen were constantly creating a variety of synthetic "environmental" drop containers, ranging from chunks of fiberglass masonry to variants of the plastic dog feces. The quest for the perfect drop device sometimes reached extremes. In one East European city, an enterprising case officer began using dead rats as receptacles for message capsules to his agents. Then he discovered that the city's hungry cats—hardly as picky as their better-fed counterparts in the West—were devouring all of his literally dead drops. He had to make other arrangements.

"That's pretty much our bag of tricks," Jacques concluded, after listing the last items in his station's tradecraft inventory.

"What about equipment delivery and refresher training?" Jacob asked.

The female officer shook her head in frustration, speaking for her colleagues. "That's the problem in Moscow, isn't it?"

We all knew a case officer often had to break free of surveillance for several minutes to several hours, so that he could safely hand over a new camera, for example, and train the agent in its use. On other occasions, the espionage equipment might simply be too large to toss or drop while under tight surveillance. In any case, agents working under the stress of spying against their own countries at such terrible risk could not remain isolated from their Western handlers forever. They needed human contact with the people they were working with in order to share experiences, voice concerns, and be assured that exfiltration was a real possibility if they sensed the net closing around them.

"But we haven't had much luck slipping their leash," Jacques admitted. On rare occasions, he said, there would be a chance break in the ring of surveillance when an officer could simply walk out of it.

"And when we're desperate," Bill Fuller added, "we give them a sacrificial lamb."

The previous summer, a case officer near the end of his tour had managed to escape for a few hours to deliver a piece of equipment to a remote dead drop in the outer suburbs. But he had been forced to engage in a provocative action when he realized he was being followed, bailing out of his vehicle, ducking into a nearby Metro station, and losing himself in the crowd.

"He made his drop," Jacques said. "But we couldn't use him for any street work after that."

"That's too high a price to pay to conduct routine operations," Jacob said, speaking for all of us.

"Well, gentlemen," Jacques said with a grin, "that's why we've invited you OTS geniuses here to help solve our problems."

Jacob laid out the key points of our survey to date, which had focused not so much on disguises as on the predictable behavior of the KGB surveillance teams. "Once a rabbit goes missing, they must have an established time limit to find him before sounding an all-points bulletin."

"Whatever that grace period is," I continued, "it's a vital piece of information for us. If we can help you slip free of surveillance without the KGB knowing you're gone, we'll have found the Silver Bullet."

Jacques raised his coffee mug. "Here's to the Silver Bullet."

JACOB AND I were at our table on the edge of the Praga dance floor, pondering Jacques's words over Armenian cognac and Cuban cigars as the band charged through Glenn Miller to Count Basie and the *nomenklatura* continued to bop and jive.

I was startled by our waiter, who was seating someone else at our table, a common practice in Eastern Europe. A quick side glance revealed a rather striking man taking a chair across from me. As he turned to face the dancers, another waiter served him a decanter of cognac and a snifter.

Although the man seemed totally absorbed in the music, there was something about him that made me uneasy. I leaned toward Jacob and nodded toward a dancer far to our right, but cast my eyes left toward the new man at our table and spoke softly. "Have you looked at this guy?"

Jacob glanced toward the dance floor, dipped his head in a brief nod, and drew back his lips in a forced smile.

Everything about the new arrival was too refined, which made him all the more menacing. His gunmetal-gray suit was hand tailored and obviously from the West. His tie was of tasteful Italian silk, his shoes gleaming kidskin. The lapel pin on his suit jacket was a familiar patriotic Soviet symbol. But I noticed that the rim, instead of being the normal brass and enamel found in tourist kiosk pins, was thin gold, surrounding a convex crystal circle etched with a portrait of Lenin against red metallic foil. The pin, like the man, was somehow alarming. They were both too finely honed to be wholesome.

After placing a box of Dunhill cigarettes and a solid-gold Dunhill lighter on the table, the man glanced at his gold Rolex Oyster watch, then reached over to shift the position of an ashtray. He knocked over my glass of cognac, but almost in the same motion, set the glass upright and refilled it from his own decanter. As the band broke into a frenetic rendition of "Take the 'A' Train," I realized that not once since sitting at the table had the man even glanced in our direction.

Now I was in a quandary. I was all too aware of incidents where Western embassy staff had been drugged in Moscow bars and restaurants to be placed in compromising positions with Soviet prostitutes or "treated" for food poisoning at specialized clinics. If I drank the cognac, became violently ill, and found myself in some isolated KGB hospital, what might I reveal under additional drug therapy? On the other hand, if I conspicuously avoided touching the refilled glass, I would practically be admitting I was a trained operations officer, not a naive, oblivious technician.

Without even moving his head, Jacob had observed the entire

drama. I knew he was ready to help me in any way he could. I cradled the snifter in my right hand, grasped the big cigar with my left, and turned in my chair to speak to Jacob over the blaring music. For a moment, my right hand was shielded by Jacob's torso and he jostled the snifter, spilling the contents onto the carpet. I exhaled a cloud of smoke, raised the empty glass, still hidden by my fingers, and made what I hoped was a convincing show of downing the contents.

Then Jacob took over, speaking loudly that we had to hurry to catch the midnight Red Arrow departing from the Leningradskaya station. With lightning speed, he paid the bill, retrieved our coats and shapkas, and ventured out into the frosty night.

Just after two that morning, as the train lunged north through the frozen swamps and taiga, Jacob made an urgent call on the burly matron of our comfortable first-class car to report his friend was sick. She offered tea, apparently heartfelt sympathy, but little practical help. *"Auf Leningrad,"* she said in broken German, *"viele guten Doktor sind."*

We settled back into our seats, content that we had played our part well in the masquerade. We hoped the matron would dutifully pass the message up the KGB pipeline that the American embassy lackey dining at Praga had been sick as a goat all night on the train ride to Leningrad.

My first operational excursion to the Soviet Union had come to a satisfying conclusion. But the real challenge would be that summer, when Jacob and I returned to test the boundaries of the Moscow rules in search of that elusive Silver Bullet.

Moscow, July 1976 ■ Jacques Dumas had not been idle in our absence. As the chill Moscow spring turned to summer, Jacques began to challenge the Moscow rules. His ultimate goal was to find a way to break free of surveillance at will, without jeopardizing agents or having to face retribution from the Seventh Chief Directorate.

As we had all agreed in February, adhering to nonthreatening behavior patterns was the key to finding the Silver Bullet. Because he was a runner, Jacques could enjoy the distance and limited freedom of discreet surveillance, as long as his pace and route were designed so as not to arouse suspicion. He was able to develop a number of sites in parks and along the Moskva River embankments suitable for placing signals and emptying one-time drops. Still, he was not comfortable making a drop while out on a run.

Once he'd established a pattern of running earlier in the morning to greet the summer dawn, he decided to push the envelope. He rose before daylight on a clear Sunday morning, pulled on his lightweight running suit over slacks and a polo shirt, and put on his ancient running shoes. Then he taped a pair of loafers under his armpits and donned a loose windbreaker.

The mili-man in the guard kiosk hardly glanced at Jacques as he trotted out of the diplomatic compound and down the sloping boulevard toward the river. No one followed him through the empty streets, and by the time he reached the Smolenskaya Embankment, Jacques was reasonably certain he was alone. Just to be sure, he'd chosen this route, knowing the pavement was blocked by the excavation trench around a broken sewer, forcing him to take narrow, meandering side streets en route to Skver Devich'ye Park.

Confident that he was indeed alone, Jacques crouched in an alley, took off his sweat-soaked running suit and shoes, and put on his loafers. He threw the running gear into a trash can and emerged onto the awakening street to embark on the second phase of his operation.

"What a feeling!" Jacques told us later that summer, when we were summoned back to Moscow. "I immediately understood the enormous potential of breaking free. I was *invisible*."

On his return to the apartment that morning, Jacques made a point of strolling down Kutuzovsky Prospekt from the north, not from the direction of the river, clutching a string bag with a bottle of fresh milk and a loaf of bread from the state *gastronom* on Kalinina Prospekt. The KGB knew our commissary was closed on Sundays, so his sudden appearance on the sidewalk with these groceries was plausible.

The Moscow office also grappled with an urgent need to free an officer from surveillance long enough to meet a potential volunteer. While this was always a tricky situation, it was even more complicated in Moscow. The volunteer could easily be a KGB provocateur, sent to ensnare the officer who came to the meeting. Even if the volunteer was legitimate, he or she could be under intense surveillance.

Since Jacques was nearing the end of his tour, he was assigned to be the action officer and offered a bold suggestion. There were members of the foreign community in Moscow whose work rarely took them into the city—and were therefore subjected to much less rigorous surveillance than suspected case officers—such as the "persons of little interest" Jacob and I had become in February. It had been established that this group, and their counterparts from other allied governments' communities, could attend social functions such as sports tournaments, picnics, and the occasional night on the town, without arousing KGB suspicion.

Jacques saw an opportunity to develop a novel variation on Moscow tradecraft. If he could somehow imitate the pattern and profile of the "little interest" group, surveillance teams might let him slip through their net. It was a radical notion, but worth a try.

The operation, which would eventually be referred to as "CLOAK," began with the open discussion of dinner plans at the Ukrania Hotel's hard-currency restaurant. To make sure that they would be seated at a booth with a window overlooking the river in the towering Stalinist building, the cover group had relied on the UPDK to call for reservations, thus making the KGB privy to the plans. In conversations certain to be picked up by hidden microphones, they spoke of the upcoming dinner on their office phones and in their apartments.

When one of the dinner group, "Len," drove his car into the courtyard on the night of the operation, the snoops lingering nearby paid scant attention. As the passengers left the apartment complex to enter the car, another of them, "Niles," patted his suit jacket, mumbled that he'd forgotten his "damn glasses" and disappeared into the apartment doorway, creating a brief diversion, during which Jacques, ostensibly an unexpected fifth member of the group, and wearing a disguise, slipped into the backseat of the big Olds sedan. Moments later, Niles returned sheepishly, brandishing his eyeglass case. The surveillance team in the outer courtyard and the street did not notice that there were now five low-ranking Americans in the car, not just four.

Len drove a fairly provocative SDR (surveillance detection route), turning sharply off Smolensky Bul'var onto Shchukina, then getting "lost" in a side street near the Mexican embassy. Confident that no one was following, Jacques removed the disguise, and when Len stopped near a Metro station, he left the car dressed as a Russian worker in a cloth cap and rust-stained overalls.

While the others headed to their dinner, Jacques kept his personal meeting with the volunteer. Once again, he had been virtually invisible to the Committee for State Security. Resuming his previous identity as one of the members of the dinner party, Jacques rejoined the other four at a pickup spot. He returned as he had left, undetected by his surveillance team, who assumed he was simply working late at the embassy.

In his postaction report cable, Jacques requested that I return to Moscow to help refine disguise methods. When I arrived that summer, he exclaimed with characteristic enthusiasm, "We've almost got our hands on the Silver Bullet, Tony."

As he described it, CLOAK closely paralleled a tactic I had been developing for an Agency office in Eastern Europe, so I was aware of the potential in this type of deception operation. But I cautioned Jacques and the other officers not to expect too much from techniques that simply altered physical appearance. "It's not the quality of the disguise that matters, but the quality of the operation," I said.

They were grateful for any help I could provide. Their optimism happened to coincide with the sudden reappearance of a potentially valuable agent, "TRINITY."

This Russian official had been recruited while serving at a large Soviet embassy in the West. Like the famous Colonel Oleg Penkovsky, TRINITY had acted from mixed motives: fervent anti-Communism, personal grievances against his corrupt superiors, and more practical considerations. He knew that if he delivered valuable intelligence to the CIA, he'd be paid well and eventually exfiltrated to live in secure and comfortable retirement in America. Therefore, he worked hard in his intense training to master the demanding subtleties of tradecraft before he was reassigned to an important ministry job in Moscow. He understood the relentless scrutiny he would be under while trying to conduct clandestine espionage close to the Kremlin walls. Once he went operational in Moscow, his survival would depend on how well he had grasped the essentials of the training.

With TRINITY's overseas tour coming to an end, his CIA handlers suggested an introduction to an American "friend" soon to be stationed in Moscow. Jacques was dispatched from Washington to meet the Russian agent and was presented as his new case officer—in effect, a living, breathing recognition signal that the KGB could never replicate to entrap TRINITY. Both men could then travel to the Soviet Union, expecting to reestablish contact safely and easily.

Jacques had been gratified when the agent had followed instructions, laying down several clandestine signals indicating that he had in fact returned to Moscow. Then, nothing happened. Six months passed with no more signals from TRINITY. He did not respond to one-way radio instructions. A full year went by.

"We were convinced we'd lost him," Jacques later told me.

By the summer of 1976, he feared that this seemingly well-motivated and potentially important agent had either been compromised by KGB counterintelligence or had become too frightened to resume communication with the Americans. For security reasons, it was too late to respond to any signals that might now be sent; they had "expired," and would have to be considered KGB entrapment if they suddenly appeared. The Moscow office wrote TRINITY off as just another source who had found the environment back home too hostile for comfort or had been rolled up.

One warm June afternoon in Krymskaya Square, Jacques spotted a man bearing an unusually strong resemblance to TRINITY among the throng of office workers trudging toward the Metro escalators. In spite of the presence of discreet foot surveillance around him, Jacques moved closer to the other man. It was TRINITY, he thought with a start, certain that the other man had also recognized him.

With Headquarters' permission, the Moscow office reactivated

TRINITY's original signals. Two days later, confirmation signals appeared at the prearranged sites. Tenuous two-way communications were reestablished, leading to the construction of a risky plan for picking up a drop from TRINITY. Everyone in the Moscow CIA office was involved in one way or another in the complex operation to service this drop. Again, Jacques, within weeks of his scheduled departure date, was sent to pick it up. The package—the crust of a sandwich wrapped in newspaper—turned out to harbor a roll of Russian 35mm film.

The Moscow office did not want to risk the possibility that the film might require special processing, so one of our officers, Nikolai, was told to hand carry it to Washington to be developed by OTS experts. Although he was not privy to the final results, he did see the first photo prints floating in the washing tray of a secure Technical Services darkroom. They were, apparently, official Soviet documents.

Washington kept silent on the value of TRINITY's initial product, but there was now a decided urgency in the tone of communications from Langley, which Bill and Jacques had never witnessed before.

"We knew something very big was happening," Jacques recalled.

On the day Jacob and I returned to Moscow, still under our former cover, Jacques gave us a draft cable that he'd prepared for IMMEDIATE transmission to Headquarters. He was due to leave on his reassignment to Washington within weeks, but he had one essential piece of business to tend to: He proposed to use the CLOAK technique, with the help of Jacob and myself, to break free of KGB surveillance for an extended personal meeting with TRINITY. CIA Moscow had never attempted such a meeting with a key asset. Jacques was suggesting that he slip through the surveillance net, meet with TRINITY for several hours, then return without leaving any trace of having been gone from the compound.

Jacob and I faced a vexing problem. Since February, my Technical

Services disguise team, supported by Jerome Calloway and other contractors, had been slaving away to prepare materials that would alter
the appearances of the Moscow contingent, should the CLOAK technique become a viable tradecraft option. However, we had nothing
ready for Jacques himself because of his imminent departure date. I
promptly wrote my own IMMEDIATE cable to follow Jacques's message,
requesting the OTS disguise team to get to work right away on disguise
materials that would allow Jacques to negotiate the streets of Moscow
unrecognized. Time was limited, and there was already an almost un-
manageable list of tasks to accomplish while in Moscow. We welcomed
the prospect of using the time while waiting for Headquarters and
TRINITY to respond.

We knew that we'd need something more sophisticated than
Jacques's earlier rudimentary disguise. Three members of the original
dinner party had either been reassigned or were on summer vacation,
so we had to bring plausible substitutes on board. One of the new members, "Roy," was already rehearsing for his role as Niles's replacement.
Len was still available, along with his hulking Olds 88. A third viable
candidate, "Jerry," volunteered for the operation, as did several other
brave souls from the nondiplomatic contingent.

We held our first "lineup" two nights later to compare the physical
appearance of these volunteers, discuss their recent surveillance patterns, and consider possible covers for the action. Roy and Jerry
emerged as the best candidates for the diversion. While keeping options
open for a variety of contingencies, we immediately started to develop
distinctive patterns for the two new volunteers to follow outside the
compound. We wanted the KGB to easily recognize them from a distance, while concluding with some certainty that they were not interesting from an operational standpoint.

In my cable to Headquarters, I requested priority delivery of an entirely new wardrobe for Roy that would make him more visible, both in the compound and on the street. Jerry already had suitable clothes that helped form part of his naturally distinctive profile.

It was essential to contact TRINITY without risking signal marks, drops, or tosses. A shortwave voice message was broadcast from outside the Soviet Union on a secret schedule in encrypted number groups, keyed to TRINITY's original one-time pads and proposing the landmark meeting with Jacques. TRINITY was instructed to make a timed drop at a secure site—a message in secret writing, stating his response to the proposal.

Bill, Jacques, and their CIA contingent worked tirelessly over the next week preparing for the operation. They selected backup signal and drop sites, then scoured the city to find an ideal secure meeting location. The materials for Jacques's street disguise arrived, and Jacob and I practically had to chase him through the more secure sections of the Station for a fitting, he was so frantic with activity. Meanwhile, we had to supervise the profile training for the other volunteers.

Roy wore a "cowboy" outfit to an embassy barbecue, the Friday night happy hour, poker games, and on short walks to shops along the Garden Ring. OTS had provided him with a creamy white Stetson hat, western shirts covered with mother-of-pearl buttons, russet jeans, and Tony Lama lizard-skin boots. To this outlandish costume, he added a pair of wide, mod-Italian sunglasses.

Jerry, aka "Big Jer," who weighed nearly 250 pounds and stood over six feet, sported stylish long hair and a neatly trimmed beard. We added a sky-blue linen sportscoat and beige linen trousers, rakish sunglasses, and a wide-brimmed Panama hat, appropriate for the midsummer heat wave gripping the city. We also backed up our wardrobe with bright pastel leisure suits that looked like *Saturday Night Fever* costumes which most Russian males over the age of sixteen secretly coveted at the time.

We decided that Jerry and his wife, "Laura," an attractive young woman with blue eyes and long blond hair, should take a weekend shopping trip to Helsinki, so that they could appear in Moscow with their dazzling new "disco" clothes. We also needed a plausible reason for Laura to engage in a new pattern of driving their car through Moscow's fast, confusing traffic in case we needed her immediate assistance at the last minute.

"Twisted your knee playing tennis up in Helsinki," Jacob suggested to Jerry. "You can start using a cane when you come back. That'll give Laura a good reason to drive you around town."

The use of the cane prop would also create a visual memory of Jerry hobbling around the courtyard even before any CLOAK outing took place. We figured that the nearby surveillance teams, being human, might appreciate watching a sexy woman like Laura fetching the family car from the parking area in front of the compound, then driving into the courtyard to pick up her crippled husband.

We were all working diligently on setting the stage with appropriate illusions that would seduce our KGB audience into "seeing" what we wanted them to see as CLOAK went into effect. First, we had to develop a safe cover action for the drop pickup. A home screening of *Jaws* was deemed ideal, since movies were a popular way of breaking the monotony of social life in Moscow. Several couples planned to watch the film at the apartment of a young case officer. The snoops were also drawn seamlessly into this event, with the host insisting that the 16mm projector and a spare projector lamp be delivered to his apartment and tested.

The evening began with an abundance of drinks and lively commentary on the realism of the great white shark. Suddenly the film broke. While one of the less technophobic husbands lugged the reel over to the audiovisual tech shop in the embassy, the host announced he was dashing out to pick up a pack of his favorite Russian cigarettes, the "stinkpots" that his wife found difficult to tolerate.

"Smoke them before you get home," she chided as he left.

The officer strode out of the courtyard archway and wandered south down Tchaikovsky Street, confident he was not being followed, since he had often been observed heading toward the tobacco kiosk, standing on a corner of Kalinina Prospekt. He was well aware that KGB surveillance teams could observe him, both coming from and going to his apartment, *except* for a crucial thirty-second period, when he disappeared into a pedestrian underpass.

Here was the site where TRINITY would make his timed drop. With a casual sweep of his hand, the officer reached into a trash can, retrieved the message container, and placed it in a dark cloth bag. Two minutes later, he was poised in front of the kiosk, chatting amicably with the vendor in broken Russian. He puffed merrily on his cigarette as he strolled back to the American compound.

The next morning he came to the office with a crushed can, still dripping with used engine oil—the concealment device for TRINITY's drop. Anyone who happened to pick up the can would have to immediately drop it or risk ruining his clothes.

When we gingerly cut the can open, we found a knotted condom enclosing three small objects and a scrap of paper. The objects turned out to be test cassettes from an advanced subminiature spy camera, which had been delivered earlier to TRINITY so he could photograph documents in his ministry and avoid the risk of smuggling classified material to his apartment. Nikolai immediately prepared the cassettes for delivery to Headquarters.

We then turned our attention to the scrap of paper. It was ragged, covered with colored-pencil scrawls. If it was found, most people would assume the paper was the product of some child playing at the kitchen table. But we understood that secret writing lay under the colored scribbles. What was TRINITY's message? we wondered. Would there be a meeting? We held our breath as Nikolai carefully carried the sheet into his tiny darkroom.

Twenty minutes later, he emerged, holding the scrap of paper. But his tightly drawn lips and pallor revealed his profound distress. "Some of that oil from the can leaked onto the paper," he said. "The message is just a bunch of dark smudges."

TRINITY's answer, which we had worked so hard to obtain, had been obliterated. We were all horrified, but Jacques looked particularly devastated. After all that meticulous labor, Murphy's Law had gone into effect.

"Let me take a look at the original message," I asked Nikolai.

Using forceps, I held the paper up against a lab lamp. The smudges were indeed illegible. But I could see that there were other problems: The paper was clay-coated, the type often used for color separations in quality periodicals but a poor choice for secret writing—the microscopic clay particles tended to bleed into the invisible ink during the development process. It also appeared that he had used his own lowercase, cursive Cyrillic instead of the block letters we preferred.

But I had a hunch we might be able to salvage something: The dark blotches reminded me of similar secret writing messages in Asian languages I had encountered in the Far East that had also been distorted.

I had been able to enhance those "ruined" messages enough for our translators to distinguish the complex characters.

"Let me have a sheet of transparent filter material," I asked Nikolai.

I slid the material between the secret message and the glass plate of the office Xerox machine. To my dismay, the blotches on the resulting copy were still indecipherable.

I turned to Jacques. "Do we have any examples of TRINITY's handwriting around?"

A rapid search through the files yielded over one hundred pages of handwritten Russian narrative, which TRINITY had produced for an autobiography while he was still in the West.

"I'm going to need a quiet place to work," I said, gathering all the materials together.

For the next two days, my skills as a forger came into play. Working in a tiny cubicle, I forged sample phrases in Cyrillic by comparing the letter formations on TRINITY's handwritten pages to the shapes of the blotches on the message. Every hour or two, Jacques came to check my progress and coached me on the nuances of cursive Cyrillic.

I had almost completed my fair copy of these phrases, which, like the diary entries I had forged in Vientiane, were totally meaningless to me, when Jacques entered the office and peered over my shoulder.

"My God," he whispered, "I can read it: 'I accept your proposal for a meeting . . . if you feel it is safe to do this . . . Wish us good luck . . .'"

After Headquarters authorized a personal meeting, the newly formed CLOAK team erupted in a quiet celebration. We were back in business and the operation would go forward.

We conducted a final dress rehearsal late on a Thursday afternoon in August. The real outing was scheduled for the next afternoon, with the cover action as another dinner. This feast would be held at the National Hotel, where Roy, Big Jerry, and Laura had become regulars on weekends. By sheer coincidence, the embassy had scheduled a large reception at the same time, so we knew the courtyard would be teeming with other diplomats and official Soviet guests. With so many high-ranking foreigners arriving that Friday night, the surveillance near the embassy would be under a great deal of stress trying to keep track of priority targets, almost certain to dismiss "persons of little interest" such as ourselves.

Just after six P.M. on Friday, Jacques and his wife, Suzette, crossed the courtyard and headed for the community center, where they frequently enjoyed the happy hour. In the apartment block, Roy dressed for the dinner, then called Jerry and Laura to remind them to stop by his place for a drink before they all went to the restaurant. I was stashed in the tiny maid's room of Roy's apartment, an airless box off an inner hall, relatively safe from KGB technical surveillance.

Jacques and Suzette worked their way slowly through the happy hour crowds, chatting with friends as they approached the door at the back of the room. In a flash they were gone, climbing the dim, narrow staircase that led to the apartment block. Suzette continued to a secure area, while Jacques tiptoed through Roy's door and into the maid's room, where I waited anxiously.

All my materials were spread out on a table so that I could outfit Jacques in a matter of minutes without saying a word. He was to assume an ordinary Muscovite working-class persona so he could move freely on the street once he had broken surveillance and bailed out of Jerry's car. He donned a faded, zipper-front sweater and a pair of frayed, "high-water" slacks. Laura would carry Jacques's drab, mismatched suit jacket and scuffed shoes in the purse she'd brought back from Helsinki. I helped Jacques slip into a detachable, muted pastel leisure suit that would pass almost unnoticed amidst the flashier colors of the dinner group, then focused on the facial attributes of his Russian persona. Working quickly with some of Jerome Calloway's best FINESSE material, I widened the tip of Jacques's nose, placed a light brown wig over his darker brown hair and finally managed to apply a blond mustache on his face, slippery with sweat.

As we made our final adjustments, I could hear the dinner party chatting pleasantly in Roy's living room. We were minutes away from the deadline, when a diversion group would drive into the courtyard, partially obscuring Jerry and Laura's car. Laura departed on cue to fetch the car, quickly followed by Roy and Jerry. Jacques stared tensely at the second hand of his Russian watch before squeezing my shoulder gratefully and slipping out the door to take the staircase down to the courtyard.

I sat down to wait again, knowing I had to remain completely silent. Only when I gasped for air did I realize that I had been holding my breath as the CLOAK team departed. Meanwhile, Jacob was sweating out the operation in a secure area, wearing headphones as he tweaked an advanced radio scanner. He logged KGB transmissions as Jacques and the dinner party piled into Jerry's car in the courtyard and drove out through the archway. The surveillance channels had been alive with staccato bursts for an hour as guests arrived for the reception, trailing their own dedicated teams of observers, who were now installed in hidden posts throughout the neighborhood.

Much to our relief, Jacques and the dinner trio returned on schedule, just before midnight. When Jacob and I joined the Saturday morning debriefing, the office was quietly buzzing.

"Everything worked exactly as planned," Jacques said, beaming with enthusiasm. "It was a textbook operation."

He had spent almost two hours with TRINITY, walking along the dark paths of Fili Park, southwest of central Moscow, and across the river. The other three had proceeded to the restaurant for a luxuriously long and festive dinner, then picked up Jacques without incident at the predetermined spot. As Laura had driven up the Rostovskaya Embankment, Jacques had maneuvered the Velcro tabs so that he was enveloped once more in the oversize leisure suit.

"We had a lot of catching up to do," Jacques said. "TRINITY needed to look into a human face, hear a real voice, and know we were still standing behind him."

A vexing mystery had also been solved: TRINITY had been very ill in the hospital for most of that silent one-year period. Jacques had assured the agent that the CIA would continue to meet its obligations and had in fact deposited monthly salary checks toward TRINITY's retirement in the West in an Agency escrow account, which had already yielded a tantalizing sum. In return, TRINITY demonstrated his good faith by passing Jacques thirty-four pages of photocopied documents and a staggering nine rolls of 35mm film.

Jacob and I immediately pitched in to help Nikolai develop the film. With the prints processed and hanging on the drying line, Bill Fuller inspected the evening's "take" and concluded that we had in TRINITY the equivalent of a highly placed penetration agent in our own National Security Council. Without arousing suspicion, TRINITY could delve into the most sensitive areas of Soviet policy.

Jacob and I knew we had participated in a watershed operational success. We had supported the CIA station as it rewrote its Moscow modus operandi, employed against the omnipresent KGB security forces. Within weeks, the CIA began to mine the Soviet Union's most confidential vaults in an intelligence haul that produced priceless information about Soviet intentions in the SALT I (strategic arms limitations) talks and secret parallel weapons development programs.

Our disguise team soon became very adept at preparing and deploying a quick-reaction Silver Bullet system for a fast-breaking outing. Disguise officers now rotated regularly between Headquarters and Moscow to exploit these opportunities. We also trained case officers and other Agency employees assigned to the Soviet Union in these new techniques, pitting case officers against the best CIA surveillance teams in realistic war games. What had begun as Jacques's desperate attempt to break free for a few moments to meet a volunteer on the street had evolved into one of the Agency's most guarded secrets, a critical technique reserved exclusively for key operations in Moscow.

TEN MONTHS LATER, on July 15, 1977, Jacques called me. He had been reassigned to Headquarters late the previous year.

"Can you come to my office at once?" he asked, his voice strained.

Inviting me to have a seat beside his desk, Jacques was pale and obviously shaken. He slid his chair close to mine and spoke painfully. "I'm sorry to have to tell you, Tony, but TRINITY is probably dead."

I was stunned for a moment, but soon recovered. "What happened?" I asked.

"We're not sure yet. One of our case officers was ambushed last night servicing a drop to TRINITY."

The case officer was "Mary Peters," one of the best and most elusive operatives under cover of the "little interest" group. For almost a year, she had serviced TRINITY's signal and drop sites without raising the slightest suspicion.

"Is she all right?"

"She's shaken," Jacques said angrily. "They roughed her up pretty bad, and she fought back, apparently hurting a couple of those thugs. Then they threw her in Lubyanka and made all sorts of threats. But the embassy consul finally sprung her. The Sovs have PNG'd her."

I sagged in my chair at this terrible news, which was the last thing any of us wanted to hear about an agent-in-place. Beyond the obvious tragedy of the situation, we were now mired in uncertainty. Had a tradecraft mistake led to TRINITY's exposure? In every case, we had tried to err on the side of caution, keeping an agent operational only as long as it seemed safe. Our exfiltration ops plan for TRINITY was already well advanced, but now it was no use.

Because the KGB had ambushed Mary Peters at a drop site, it was clear that they had managed to extract information from TRINITY. How long had they been controlling him? How much of his product was valid? Such questions could not easily be answered. But I did know from the Directorate of Operations that most of TRINITY's remarkable flood of intelligence had been of unprecedented sterling quality, so his compromise must have been recent.

Over the coming weeks, we all watched tensely for Soviet government reaction. The KGB was always very careful about making a plausible espionage case against a government official. Once they had enough evidence for a convincing public trial, however, they proceeded quickly with the process, followed by a death sentence.

While waiting for TRINITY's trial and inevitable sentence, the DO and OTS performed an urgent damage control assessment. What tradecraft secrets had been compromised? Exactly what could TRINITY have told the KGB about our people and methods? What sensitive materials and equipment had we given him? Jacques and I both agreed that

TRINITY did not know that we had been using the CLOAK technique, so the Silver Bullet was still intact.

But now the KGB probably understood that Mary Peters had led a very disciplined lifestyle, with a low profile and nonthreatening behavior pattern. Her only operational task had been to maintain impersonal commo with TRINITY when she was confident no surveillance was present, and she had always conducted a long, complex cleansing run to be certain that she was alone. Now we had to dissect everything about her persona, patterns, and outings to make certain we hadn't neglected some aspect of tradecraft that might have compromised TRINITY. It was time for a Blue Ribbon internal panel to fully review Moscow rules.

"GORE HARRINGTON," the new chief in Moscow who replaced Bill Fuller that fall, was plainly no fan of the Silver Bullet option in Moscow tradecraft. Pushing Moscow rules beyond the limits, he grumbled, might have been what had led to the failure of the TRINITY case and the death of one of our most courageous and effective penetration agents.

As winter set in, a parallel chill gripped the Soviet–East European Division. Over the coming months seven of their agents-in-place in the Soviet Union and elsewhere were compromised. And then a suspicious fire broke out in the American embassy in Moscow; it seemed to have been rigged to give the KGB access to our most sensitive offices. Was our string of good luck in Moscow ending?

The Carter administration had been in the White House for almost a year, and the patience of the new Director of Central Intelligence, Admiral Stansfield Turner, was wearing thin when it came to Soviet operations. Because we suddenly seemed incapable of working effectively against KGB opposition, Turner considered the withdrawal of our entire contingent from Moscow.

A few months later, the panel of inquiry on the 1977 compromises released their classified findings. According to unconfirmed information provided by a KGB agent still in place, TRINITY had been seen photographing secret documents at his ministry. Knowing that he faced brutal interrogation, he had agreed to write a full confession. But instead, he commited suicide, using a cyanide capsule ("L" pill) he had obtained from Jacques.

The panel also found breaches in operational security by the other compromised agents. DCI Turner now seemed somewhat less likely to pull the plug on Moscow.

To me, the account of TRINITY's compromise seemed plausible. But the panel's report did not explain how the KGB counterintelligence teams had managed to ambush Mary Peters the night she went to service TRINITY's drop site. For years, I simply assumed that he had folded under interrogation and revealed his tradecraft repertoire before killing himself.

I was wrong.

In 1998, a Russian writer named Sergei Gorlenko published a somewhat questionable account of the TRINITY case on an Internet website (www.intelligence.ru/english/public/n0001). Gorlenko apparently had access to detailed files on TRINITY's arrest and Mary Peters's subsequent unmasking.

This version paralleled known events: Gorlenko confirmed that TRINITY had served at a large Soviet embassy in the West in the mid-1970s. The KGB's First Chief Directorate reportedly received information from an agent that someone in the Soviet embassy who used the sports club at the local Hilton Hotel had been recruited by the CIA. The KGB initially identified twelve possible suspects who were quickly narrowed down to three, including TRINITY. Coinciding with TRINITY's return to Moscow, Second Chief Directorate (Counterintelligence) technicians noted new clandestine coded radio broadcasts originating in Western Europe, suggesting that an agent-in-place had just been activated in the Soviet Union. Further, Gorlenko stated that "laborious analysis" determined this agent was either unmarried or divorced and maintained his own bachelor apartment.

TRINITY was the only one of the original suspects who fit this category. And his behavior gave further weight to the evidence against him. TRINITY turned down a prestigious academic position to take a lower paid, dead-end job as a clerk in his ministry. But this clerical assignment provided him access to highly sensitive Soviet national security information.

Despite this suspicion, however, Gorlenko stated that the KGB allowed TRINITY—now assigned the Russian cryptonym "AGRONOM" (Market Gardener)—to continue in his ministry position while they slowly developed their investigation. Surveillance teams discovered he was probably using his Volga sedan, often parked on certain side streets following an apparent schedule, as a signal to American case officers. He would then visit Pobedy Park in west central Moscow late in the evening, walking paths taken by known CIA officers from the Moscow station.

Based on this now overwhelming evidence, a KGB counterintelligence/surveillance team searched TRINITY's apartment while he was away on a trip in mid-1976 and installed a hidden mini-video camera. Although they were unable to unearth any obvious spy gear, TRINITY did have a transistor shortwave radio with a built-in cassette recorder, ideal for receiving coded broadcasts.

It was not until June 1977—according to Gorlenko, almost two and a half years after TRINITY had returned from abroad—that the surveillance camera revealed him removing a transparent plastic fiche of communication instructions (including coded references to the sites and schedules of timed drops), hidden in a dummy flashlight battery.

On the evening of June 21, 1977, the "capture group" from the Second and Seventh Chief Directorates arrested TRINITY in the courtyard of his apartment house. He led his captors to a garage where he had hidden his other espionage equipment. The most damning was a dead-drop container, an OTS-produced artificial cobblestone with a hollow interior.

According to Gorlenko, TRINITY "expressed a desire to write with his own hand the testimony about his spying activities." His captors granted him this wish. TRINITY picked up his Waterman fountain pen, wrote a few lines, then jammed the pen into his mouth, bit through the barrel, and fell to the floor. The KGB team was shocked. Their effort to pry the pen and its cyanide capsule from TRINITY's mouth with a wooden ruler failed. He was pronounced dead at a nearby hospital.

In this version of events, the KGB had clearly blundered in managing the investigation. They now tried to redeem themselves by capturing TRINITY's case officer. Using his cryptographic instructions, they decoded radio instructions suggesting a meeting at the "Forest" site, which they decided was Pobedy Park. They disguised a KGB surveillance officer as TRINITY and had him drive the agent's Volga to a signal site, thus agreeing to a meeting in the park. But the surveillance teams located no American officer in the park on the night of the meeting.

(They were unaware that Mary Peters, TRINITY's street contact officer, had in fact been in the park that night.)

Undaunted, the KGB pressed ahead, trying to trap TRINITY's case officer. Another broadcast recommended a dead-drop exchange at a site bearing the code word "Setun," the Krasnokaluzhsky railway bridge across the Moscow River west of the city's center. Once more, the agent had to confirm the meeting at a separate signal site, this time by placing a lipstick smear on a lamp pole bearing a triangular children-crossing sign on Krupskaya Street.

But the KGB were surprised when Mary Peters was observed slowly cruising by the lamp pole. Her cover, deep in the "other" group of uninteresting drudges in the embassy, had been so effective that she had never been assigned a street surveillance team. That changed immediately.

Saturation surveillance on Mary observed her reading the lamp pole signal on the evening of July 14. The meeting on the railway bridge was scheduled for the next night. The KGB had prepared a rough welcome for her. The bridge, a prerevolutionary construction of rusty trestles and sooty stone towers carried double railway tracks across the river, as well as a narrow steel-plate walkway. It was not a pleasant place for a stroll on a dark night, but it did provide good visibility to detect close-in surveillance and also a convenient dumping ground for incriminating material: the Moscow River.

To ambush Mary, the KGB cut a hatch in the warped steel plates of the walkway and connected it to the catwalk below with a ladder. A small surveillance team would be lurking there when Mary made her way overhead on the pedestrian passage.

The night of July 15, Mary was observed leaving the American housing compound in her car "with her hair carefully styled and with a fashionable dress on." But when she reached the cinema at the Rossiya Hotel, she climbed from her car wearing a dark blouse and slacks, and her hair was down. She reached the railway bridge by Metro, scanned the area for surveillance, and then paused, making certain she was alone. Although the hidden surveillance teams had no way of knowing, she was listening carefully for their scrambled voice-code radio messages, using an Agency-adapted earpiece and a special receiver strapped beneath her blouse.

Hearing none of the usual cryptic chatter of the "foot" and car teams, Mary mounted the echoing steel staircase to the bridge and made her way along the narrow walkway beside the tracks. Just as she entered the arched passage through the stone-block tower, the hatch on the walkway clanged open and the KGB arrest team sprang out to seize her. While she screamed loudly in Russian that this was a "provocation," reinforcements pounded toward her from either end of the bridge.

Mary knew, of course, that they had her cold when they wrested from her grip the dummy cobblestone containing cash, mini-cameras, microfilm, and coded messages for TRINITY. But she resisted vigorously, hoping to alert the agent if somehow he might still be approaching the bridge.

Her effort cost her some bruised ribs, a wrenched arm, and a black eye. By the next morning, the American consul had arranged her release from the Lubyanka Prison. She was declared persona non grata and expelled a day later.

Reading Gorlenko's account, I detected a mix of previously secret fact and likely falsehood. First, the revisionism: The case history was written by the KGB, which had the unenviable task of publicizing their success in capturing Mary Peters, while simultaneously trying to deflect attention from TRINITY's considerable success as an espionage agent.

Further, the former KGB also had to explain how one of the Soviet government's most promising young officials had been recruited at a large embassy in the West well-staffed by counterintelligence experts.

Gorlenko's version of the truth reflects these difficulties. To overcome them, the account all but ignores the year between the summers of 1976 and 1977, after Jacques reactivated TRINITY and the agent delivered his trove of extremely valuable intelligence.

So, while the step-by-step record of Mary's surveillance and capture on the bridge is completely accurate, I find it hard to believe that the Second Chief Directorate had narrowed down the list of suspected American agents to TRINITY alone by the time he returned to Moscow, then simply allowed him to continue in his sensitive position for two and a half years. And it is just not plausible that they would have given him completely free rein for the first eighteen months of this period before installing video surveillance in his small apartment. It is more likely that our agent made a tradecraft mistake, as the Agency's investigative panel indicated, and that this error led to his subsequent surveillance and capture.

In short, I think the Internet account of TRINITY's demise and Mary Peters's arrest represents an overstated historical account, diluted with some good old-fashioned bureaucratic Cover Your Ass revisionism that has survived the collapse of Communism and the Soviet empire.

BUT THE COLD WAR was still a reality in 1978 when a situation in Moscow fully restored both Turner's and Gore Harrington's confidence in our Silver Bullet techniques.

CIA had received word that an internationally prominent, English-language magazine published in the United States was about to print a story that would inadvertently cast suspicion on a retired senior Soviet official as a former secret asset of the FBI and CIA. The Chief of the Soviet–East European Division appealed to the magazine's editor to kill the story, but this was impossible: The issue was already printed. Publication of the article, however, would place our loyal former agent in extreme jeopardy.

Since the retired agent was no longer in contact with the Moscow office, it was imperative that one of our officers slip free of surveillance to warn the Russian and offer him the option of emergency exfiltration before the story broke. But those who knew this old official, whose health was failing, feared he would refuse the offer, for he loved his Russian homeland as passionately as he despised the ruling Soviet system.

On the chance that the former agent would accept the offer, my Technical Services group went ahead with preparations for the very first clandestine exfiltration from Moscow. Gore Harrington, still disdainful of CLOAK, opted to be the action officer who would contact the agent. He was confident that he could elude his watchers without resorting to the Silver Bullet option.

But I had other problems to worry about. Since I had been promoted from Chief of Disguise to Deputy of Authentication the year before, it was my responsibility to oversee and plan every detail of the risky exfiltration, including all the document and disguise options for the subject and the American team. Because he was well known, transporting the agent through any airport in Moscow or elsewhere in the Soviet Union was not feasible, but we had a fair chance of exfiltrating him by land or sea to another Soviet bloc country, where we could disguise and move him safely to the West. I marshaled my troops, both at Headquarters and in Europe, urgently mounting parallel probes along train and ferry routes, so that we could update our information on controls, with a focus on how the Soviet bloc security system treated a traveler with a particular third-country alias and cover. Jacob's capable team responded immediately, and the first probes were dispatched. When I found that we had no recent passport photo of the agent, my shop produced a suitable stand-in. We disguised him to closely resemble our subject and photographed him in a variety of poses for his exfiltration passport and other documents. The project consumed us as we worked mind-numbing eighteen-hour days.

Early one morning several days later, my secretary brought me a restricted-handling Secret cable from Gore Harrington. Despite a concerted effort, he had been unable to break free of KGB surveillance. As a last-ditch attempt, however, he had turned to the Moscow CLOAK team we had trained and reluctantly chose an "off-the-shelf" Silver Bullet option, which we had prepared for him despite his earlier resistance. On that successful outing, Gore Harrington, like his colleagues before him, had discovered the intoxicating power of invisibility.

Unfortunately, his clandestine meeting with the retired agent was less productive. The old man's emotional bond to his native Russian soil was too powerful for him to flee into exile. He thanked Gore for all we had done, but bravely chose to accept his fate.

"I will die here as I have tried to live," he told Gore, "with dignity."

Reviewing this entire episode, DCI Turner declared the use of the Silver Bullet technique "a fine piece of work." Gore went even further, stating that the CLOAK–Silver Bullet combination was his "most valuable capability" and should be maintained in Moscow as a fail-safe tradecraft option.

What had begun as an exercise in frustration had ended with one of the most innovative espionage tradecraft systems in our history. The widespread acceptance of these techniques ushered in an unusually productive period for the Moscow station, with one success after another. These fruitful years would last well into the next decade, until we suffered betrayal at the hands of one of our own, an embittered former career trainee in the Directorate of Operations named Edward L. Howard.

But before I would confront internal treachery, I found myself on another geopolitical fault line—revolutionary Iran.

9 ■ RAPTOR in the Dark

> If plans relating to secret operations are prematurely
> divulged, the agent and all those to whom he spoke of
> them shall be put to death.
>
> —Sun Tzu

Tehran, April 1979 ■ The tall Iranian stood facing me in the dim light of the cramped bathroom, gripping a strip of flat television antenna wire with both his hands. A naked bulb attached to the twisted copper ends of the antenna dangled from his right fist, while his left hand thrust the opposite ends of the wire into an outlet beside the bathroom sink. The door was shut to prevent the light of this improvised lamp from spilling into the other rooms of the dark apartment.

It was mid-April 1979, six months into the mounting chaos of the Iranian Revolution, triggered by the Islamic fundamentalist Ayatollah Khomeini and his followers. The Shah, self-proclaimed heir to the Peacock Throne of the ancient Persian Empire, was in exile, stricken with cancer. On February 14, two weeks after Ayatollah Khomeini had returned triumphantly from his own European exile, an anti-American mob overran and briefly occupied the sprawling U.S. embassy compound on Takhat-e-Jamshid Avenue in central Tehran.

Revolutionary Guards loyal to the Kometeh (the Committee for the Revolution) expelled the mob, then used the takeover as a pretext to roam the compound themselves, ostensibly "guarding" the threatened American diplomats. But these same custodians of order prowled the streets every night, firing submachine guns into the air and chanting in Farsi, "Death to the Shah! Death to America! Death to the CIA!" Tehran had become a city ruled by gangs of well-armed zealots whose loyalty lay with a shifting alliance of Muslim clerics loosely united under Khomeini.

This escalating violence had triggered the Tehran station's urgent request for the exfiltration operation of our most valued Iranian agent (code-named "RAPTOR"), the man who now stood beside me. We were in an apartment on the third story of a nondescript building two blocks from Motahari Boulevard, where I worked as quickly as I could on the preliminary stages of RAPTOR's exfiltration disguise.

The apartment was desolate but displayed evidence that RAPTOR had been camping here for several weeks. A soiled blanket was thrown across a sofa in the adjoining bedroom, and beside the sofa stood a vintage wooden console television set. He had obviously been tinkering with it, since its tubes and wires were strewn about the floor. The sofa and television were the only pieces of furniture in the apartment. In the narrow kitchen, there was a stack of well-thumbed foreign magazines and a pile of old Farsi newspapers. A bag of rice, a sack of lentils, and a single row of canned food from the bazaar were neatly stacked on the drainboard of the kitchen sink. All the windows were covered with layers of newspaper in lieu of drapes or blinds.

"Just a few more minutes," I whispered to RAPTOR, trying to assure him that the special disguise materials on his face would soon be removed. His nose and cheeks, from his upper lip to his brow, were now covered with a material that obscured his vision. Forced to breathe through his mouth, he was clearly fighting off the claustrophobic panic that the loss of sight and normal breathing could induce.

On the other side of the sink, "Andrew" was assisting me, stirring a special adhesive under a stream of warm water from the rusted tap, while "Hal," the acting CIA chief in Tehran, sat tensely on the sofa, listening to the subdued crackle of a Motorola two-way radio held against his ear.

THIRTY MINUTES EARLIER, as a smoggy night descended on the city, Hal, Andrew, and I had completed our surveillance detection run by passing through a bustling department store on Abbasabad Avenue. After leaving the store, we had dashed across the street, dodging cars, taxis, and trucks, most of which sped through the streets without headlights. Our risky move had the effect of making any foot surveillance visible; plunging recklessly into the chaotic onslaught of traffic would also thwart any vehicle surveillance. We would never have used such a blatant SDR in a sophisticated spy capital like Moscow, but the opposition in Tehran was essentially composed of bandits, not trained counterintelligence operatives.

Certain we hadn't been followed, we made our way quickly around a corner to the side entrance of the apartment house. Two other CIA officers parked down the street saw us coming and scanned both ends of the block to confirm that we were clean before we entered the doorway. They then signaled by radio code that it was safe to proceed.

We crept up the dark stairs and found RAPTOR hiding in the shadows of the second-floor landing. He stepped into the dim light. The gaunt, lanky middle-aged man still bore the unmistakable stamp of a senior military officer, despite his ill-fitting civilian sweater and trousers.

During the previous ten years, RAPTOR had advanced swiftly in the Shah's armed services, eventually assuming a key staff position in the palace. Over that decade, he had been a prized unilateral intelligence source for the CIA. The scion of a cultured, wealthy family close to the royal court, RAPTOR had nurtured a direct connection to the Shah and had been privy to his policies during the tumultuous years preceding the Islamic Revolution. In fact, RAPTOR had been our sole "Blue Striper" (top-level) source in Iran, whose intelligence was so reliable that it was sent directly to the White House on receipt at Headquarters. But when Khomeini had declared his Islamic Republic in February, many Iranian civil officials and military officers had fled the country, fearing brutal imprisonment or execution. RAPTOR had immediately gone into hiding. He had spent the first few weeks in a relative's unheated attic, crouched beneath a tin roof on which melting snow dripped continually.

When we were safely inside the flat, RAPTOR embraced us, his eyes brimming with tears of gratitude. Darting quickly across the living room Andrew and Hal pulled open a narrow window and dropped a coiled rope to the bottom of the light shaft forty feet below, which adjoined a commercial hotel facing a busy avenue. The rope was our emergency escape route, to be used if we had to flee and the stairway was not an option. After descending the light shaft, we could enter the hotel through the laundry window and leave through a service entrance.

Watching us, RAPTOR regained his characteristic decisiveness and he took me into the bathroom to display his handiwork with the TV antenna and the forty-watt bulb. Without speaking, I laid out my materials, and Andrew joined us.

■

I TESTED THE disguise with my fingertips, but the material had not quite set. Suddenly, we heard a knocking on the front door . . . three faint taps. All of us froze. RAPTOR pulled the wires from the outlet, and I opened the bathroom door. Hal was whispering urgently into the radio. Blinded by the disguise material, RAPTOR groped his way toward the door as I led him by the hand, with Andrew and Hal close beside me.

Were we about to be caught red-handed? Andrew and Hal slipped past us to ready the rope in the open window.

The knocking persisted. With my hand guiding his, RAPTOR bent down, his mouth close to the door, and whispered in Farsi.

"Who's there?" he asked, his voice muted because of the disguise material covering half his face. It glowed weirdly in the faint light.

"It's me, Uncle," came the voice of a young boy.

It was the son of one of RAPTOR's relatives, who owned several flats in the building. Slowly, we exhaled.

"Do you need anything from the bazaar, Uncle?" the boy asked.

"No, lad. Not now. Come see me later."

"I will, Uncle." The child's footsteps faded away on the stairs.

■

RAPTOR AND I spent the next three days together at Hal's safe house apartment. But "safe" was hardly an appropriate word: In February, Hal had watched in horror as militants burst into the lobby with a .50 caliber machine gun and started firing away at the high brick walls of the nearby American embassy compound. Such unpredictable and gratuitous violence seemed inevitable in a country on the brink of anarchy. Even in the absence of organized surveillance, Revolutionary Guards, militant "students" loyal to their own mullah or ayatollah, or renegade former soldiers who had adopted Islamic zealotry, might simply decide to break down our door in search of booty and capture us in the process.

We planned to exfiltrate RAPTOR out through Mehrabad Airport right under the noses of the Kometeh security service and their armed enforcers, the Revolutionary Guards. We also decided to transform RAPTOR into an elderly Jordanian businessman, an Anglophile who favored rough tweed and had adopted the British manner of the old Trans-Jordan protectorate. This persona was chosen because RAPTOR spoke decent Arabic and could apply a British accent to the English he had learned at U.S. military schools.

On the day before the scheduled departure, we assembled all the elements of the disguise, wardrobe, and alias documentation. RAPTOR sat at the dining table of the flat in his lumpy woolen suit, carefully scanning the well-worn passport and other identity documents Andrew had provided. He looked up and smiled, transformed into an old Arab salesman who had traveled the Gulf states for decades, peddling oilfield equipment and truck parts. I could tell from his expression that we had managed to instill some trust in our subject, and I could only hope that it was warranted.

Since RAPTOR had never seen Andrew undisguised, we could use him as the spotter at the airport. His final task before boarding RAPTOR's flight would be to make a phone call from the public booth in the departure lounge and pass a "go" or "no go" signal. If RAPTOR boarded the plane safely, Andrew could then identify himself after takeoff and assume his duties as the Iranian's escort to freedom. If things did not go well, Andrew would presumably witness RAPTOR's capture and have seen where he was taken, then report to us.

RAPTOR seemed comfortable in his clothing and with his personal effects, which greatly augmented his disguise persona. I had coached him for hours on how to walk and talk, and especially on the somewhat doddering manner to employ, presenting his airline tickets, passport, and customs declaration. Andrew briefed him for over an hour on the alias documents, ostensible travel itinerary, and cover legend. RAPTOR had memorized the phone numbers of his "affiliate" offices elsewhere in the Middle East, CIA fronts already alerted to vouch for his bona fides, should Iranian officials at the airport somehow decide to call.

After RAPTOR turned in following dinner, however, I had a final private word with Hal.

"He's been a quick study and seems eager to meet the challenge," I admitted. "But I'm worried."

Three times over the past few days, RAPTOR had visibly slipped into depression. He had hunched in his chair, hands clasped between his knees, face tense. "What will happen if things go wrong at the airport?"

His greatest fear was being caught, tortured, and made to lose face in front of his captors. "You have no idea what they would *do* to someone like me," he said. "Don't you have a pill for me that I could use in such an emergency?"

I thought of TRINITY in Moscow. But I had neither the time nor the volition to make such a request of Headquarters. "You're just feeling operational nerves," I assured RAPTOR. "Everything will be fine. We've done this many times before and we'll be with you every step of the way."

As a diversion, I demonstrated my old sleight of hand with two wine corks. "This is how we will slip through the airport, just as though we were invisible."

"What's your assessment?" Hal asked after I related these incidents.

"We'll wake him up very early tomorrow and I'll make my final assessment of his mental state then."

If I decided RAPTOR was not up to passing through the airport alone, I would buck Headquarters' instructions and personally see him through the controls. I would not need to board the airplane with him, but would at least escort him safely into the departure lounge.

The next morning at three I awoke RAPTOR, now known as "Mr. Kassim." Once he'd had a shower, I worked quickly to apply the disguise. But it soon became apparent that Kassim had changed overnight in fundamental and significant ways.

"Did you sleep well?" I asked.

He merely clicked his lip, the prototypical Middle-Eastern negative response. The subtle color matching I had struggled with preparing the disguise no longer conformed with his greenish pallor. His manner was also altered. There was no mistaking the hunted look in his face.

I understood his dilemma. Before reaching senior rank, RAPTOR had been an energetic young field officer. His soldier's instinct was toward action. He preferred crossing hundreds of miles of mountains and desert on foot and shooting his way across a border, dying honorably in the action if that was God's will. But to be caught sneaking from Iran unarmed, dressed like an old man, went against the grain of a once-proud soldier. He would need help at the airport.

While Andrew prepared a light breakfast of flatbread and tea, I took Hal aside once more and informed him of my decision.

"I'm going into the airport with Kassim." Hal nodded his agreement. "I'll go ahead with Andrew to reconnoiter the terminal one last time," I added, "and confirm that the Swissair flight has landed and is scheduled to depart on time." This was definitely stretching Headquarters' instructions thin. But we were the operational officers on the scene and had to take chances or risk failure.

As we drove through the empty streets of Tehran before dawn on the crucial day, the acrid coal smoke and anti-American posters plastered on every bare wall heightened the menace of the deserted city and accentuated the gravity of our task. Driving beneath the ornate archways of Mehrabad Airport and around the circular drive we stopped near the glass-and-concrete main terminal. We were right on schedule. After Andrew parked the car and deftly removed his disguise, we strolled into the terminal. Glancing around quickly, I felt optimistic. The hall was empty except for a few drowsy militants slouched on benches and some temporary revolutionary officials sipping tea at their counters. No one seemed to pay attention to the two foreigners who had just entered the terminal. The Swissair agent confirmed that the plane was en route from Zurich and on time for both landing and departure. Andrew proceeded through immigration control with no problem. *So far, so good*, I thought.

I returned to the curb and waited for Hal and our star performer, Mr. Kassim. To avoid looking suspicious, I strolled to the dark end of the parking lot and watched the sunrise for a few moments to calm myself. As the sun lit up this promising spring day, taxis and vans rolled to the terminal to unload passengers, and soon Hal and Mr. Kassim arrived. They climbed out of the cab, and I walked up casually, offered a broad smile, and shook Mr. Kassim's hand to reassure him. I then grabbed his luggage and nodded a farewell to Hal, who would return to the safe house and await the signal from Andrew that "Mr. K" was either safely aboard the flight or had not made it—an unenviable job.

Mr. Kassim passed through the customs departure control with no problem. Andrew's documents were meticulous, and the disguise of a well-traveled, elderly Arab merchant aroused no suspicion among the amateurish revolutionary-zealot customs agents, who had been alerted to search for wealthy Iranians fleeing the Islamic Republic with gold, valuable carpets, or antiques. After Kassim checked in with Swissair and had his boarding pass, I took him as far as the immigration checkpoint and hung back as the Revolutionary Guard clerk stamped his passport.

Gripping Mr. Kassim's hand to say good-bye, I could see the same glint of animal terror in his eyes as before. He abruptly broke away and proceeded down the corridor to the departure lounge. Something was not right. Even though I should have left the airport, I decided to stay in the terminal until the Swissair flight had departed.

Twenty minutes later, I was seated in the waiting area, listening for the boarding announcement, when I looked up and saw Andrew motioning to me through the clear glass barrier on the other side of the immigration checkpoint, his face grave.

"The flight's been called," he whispered. "But I can't find Mr. K. I saw him come into the departure lounge, but he's just disappeared."

I thought fast. Could this operation be salvaged? "Go back and board the flight," I said.

I returned to the Swissair desk and told the agent I had a problem.

"My uncle is boarding your flight to Zurich, but I'm afraid I've forgotten to give him his heart medicine."

The man frowned.

"Can you escort me through Immigration so I can find him and make sure he has the medicine and knows how to take it? You see, he's an old man and very forgetful."

The airline clerk nodded sympathetically and immediately walked me through the checkpoint, then turned me loose in the departure lounge. Across the hall, Andrew was preparing to pass through the final security check and board the bus to the airplane. He gave me a shrug— he still didn't know Mr. K's whereabouts.

I scanned the wide hall. Suddenly my eye fell on the door to the men's lavatory. He had to be in there, I thought. As soon as I entered the echoing, tiled room, I noticed that one of the stalls was occupied.

"Mr. Kassim, Mr. Kassim," I called softly.

The stall door opened slightly, and a large dark eye peered anxiously back at me.

"Come on, Mr. Kassim. You'll miss your flight."

The door opened further. He was transparently shocked at seeing me.

"Come on, my friend," I said with a chuckle to reassure him. "It's time to go."

"How did you get in here?" he stammered.

Seizing him by the elbow, I thrust him from the men's room and led him across the departure room, arriving just in time for the bus. Uniformed Revolutionary Guards, who were now obviously present in the departure lounge, glanced at us curiously but did not approach.

Although he had been literally immobilized by fear, my sudden appearance had broken the grip of his terror.

Five minutes later, the clerk informed me the flight was on its way to Zurich. I placed my call to Hal, signaling that the exfiltration had been successful. That afternoon, the return cable from Andrew was welcomed by all: RAPTOR LANDED SAFELY IN HIS NEST.

Washington, November 1979 ■ In the early fall of 1979, I had been named Chief of Authentication for the Graphics and Authentication Division (GAD) of OTS. I was now responsible for disguise, false documentation, and the counterterror and counterintelligence forensic examination of questioned (possibly forged) documents and materials. After the RAPTOR operation, I made sure that we renewed our active contingency plans for agent rescues from hostile territory.

From the moment an agent came on board, GAD specialists began to prepare for his eventual exfiltration, which required the labor of hundreds of personal documentation and disguise specialists. The documents themselves were produced by graphic artists and craftsmen, whose products had to pass the most rigid scrutiny. At the same time, regional experts tailored an agent's cover legend to match his demeanor and personality. Finally, technical operations officers had to be prepared to lead the operation in the field as I had, literally shepherding the escapee through the checkpoints to freedom, making sure that he did not lose confidence at the critical moment.

The ultimate test of our abilities would come in November 1979, when radical "student" followers of the Ayatollah Khomeini seized and occupied the American embassy in Tehran. Over the years, the 444-day-long captivity of the fifty-three hostages from the American embassy in Tehran, as well as the secret rescue of their six State Department colleagues on January 28, 1980, have been the stuff of myth. The role of the CIA in general, and my participation in particular, remained secret until the CIA's fiftieth anniversary in 1997.

Jean Pelletier and Claude Adams's book, *The Canadian Caper (The True Story of Six Americans and Their Daring Escape from the Hostage Crisis in Iran)*, told only part of the story. Written soon after the crisis, it mentioned that some CIA help was provided to Canadian ambassador Kenneth Taylor in his escape plan for six American "houseguests," who were hidden at the homes of Taylor and his chief immigration officer, John Sheardown. But since then, the prevailing impression has been that it was the Canadians alone who rescued the American diplomats from the vengeful Iranian mob hunting them down, and somehow managed to lead the six diplomats through Mehrabad Airport to freedom.

I am now at liberty to tell the true story of those tense, dramatic days.

MY MEMORY OF the prolonged national emergency begins on the morning of November 4, 1979, when I entered the CIA's Foggy Bottom compound. The South Building seemed to be under a state of siege, with people striding grimly in all directions, clutching red-striped Secret files. A quick scan of the wire service tickers and cables from the field confirmed the worst. Hundreds of militant "students" shrieking, "Death to America!" had overrun the embassy compound as armed Revolutionary Guards stood by, or even aided them in their efforts. Ayatollah Khomeini quickly voiced his personal support for the unprecedented act of aggression. The presence of the deposed Shah in New York, where he was undergoing treatment for advanced lymphoma, had moved Khomeini and his followers to demand the Shah's immediate return, along with all Iranian government funds held by American financial institutions.

Within days, however, a glimmer of hope emerged from the grave cables and shocking news photos of American diplomats, bound and blindfolded, herded at gunpoint by bearded young militants before a chanting mob of thousands surrounding the embassy compound. Apparently, five Americans—men and women working in the separate consular section building at the rear of the compound—had managed to escape during a sudden rainstorm and were taken in by the Canadian embassy. The five individuals were Consul General Robert Anders; Consul Joseph Stafford and his wife, Kathleen; and a second vice consul, Mark Lijek with his wife, Cora. Agricultural Attaché Lee Schatz had been working in a nearby office when the mob attacked. Eventually he was able to join the other Americans, making a total of six.

Canadian ambassador Ken Taylor sheltered Joseph and Kathleen Stafford in his official residence in the comfortable suburb of Shemiran, while Chief Immigration Officer John Sheardown and his wife, Zena, harbored the other four American fugitives in their nearby villa. Kept out of sight of the prowling Revolutionary Guards and Kometeh plainclothes operatives, the Americans were in no immediate danger.

But any day, the unpredictable Kometeh and Revolutionary Guards might double check their head count of the embassy hostages and come up six short. This was not yet an immediate threat because the situation in Tehran remained chaotic, and the Iranians couldn't be sure precisely how many official Americans had been in the country and how many away on leave at the time of the takeover. If the hostage-takers did realize there were six American diplomats still free in Tehran, however, the Revolutionary Guards might search all diplomatic buildings, immunity be damned.

The fundamentalists in Tehran were in a frenzy of vituperation. The Ayatollah Khomeini, speaking on state television from the holy city of

Qum, had declared the occupied U.S. embassy "an espionage place" that had worked for decades against the Iranian people. Now we had reason to worry. The fact that the six houseguests had evaded the militants' initial assault—a possible indication of clandestine training—might unfairly brand the Americans as members of the despised CIA and could lead to harsh treatment if they were captured.

An equally ominous event almost went unnoticed during these chaotic days. The Kometeh's secret courts condemned over six hundred allegedly anti-Khomeini prisoners to immediate execution. For three days, they were led before firing squads in the prisons once run by the Shah's brutal secret police, SAVAK. As I read these accounts, the plight of the six Americans began to haunt me. The militants combined hatred of the West with stubborn vindictiveness. They might actually execute the six American houseguests, if they could get their hands on them.

In mid-December 1979, my division chief entered my office early on a Thursday morning and dropped the case files concerning the six Americans on my desk.

"I know you've got plenty on your plate," he said. "But this job is going to have to take an equally high precedence."

"I'll form an exfiltration planning team immediately," I promised.

In any exfiltration, failure was a catastrophe. In this case, if we were caught in an attempt to rescue the six American diplomats, our failure would receive immediate worldwide attention. Not only would it reflect poorly on the United States, President Carter, and the CIA, but it would also place the other embassy hostages in Tehran in an even more dangerous situation. The Canadians were in dire straits as well, and if our exfiltration were to unravel, we would drag their diplomats into the mess.

As usual, we were working under extreme time pressure. First, we had to evaluate the basics of the exfiltration. Fortunately, RAPTOR's rescue had given us a body of technical data on then-existing airport controls, so now we had to collect and analyze current intelligence on Iranian Customs and Immigration, particularly concerning foreigners. To so do meant continuing to support the infiltration and exfiltration of a few officers and third-country agents traveling in and out of Iran on intelligence-gathering and hostage-rescue planning operations. We could turn their information to our own purposes.

We were most concerned about the positive immigration exit controls that dated back to the days of SAVAK's iron efficiency. Iranian Immigration still used the two-sheet embarkation/disembarkation form, printed on No Carbon Required (NCR) paper, which foreign travelers had to complete on arrival. The original white copy remained with Immigration, and the traveler retained the second, yellow copy in his passport. Upon departure, the traveler presented this copy at the immigration exit control, where, in theory, the clerk compared the two copies of the form to verify that the traveler was departing before his visa expired.

This control procedure was similar to what we had encountered during the NESTOR exfiltration. But while Jacob and NESTOR had been able to bluff their way through Customs without their currency declaration sheets, it was foolish to think that six nervous American fugitives could brazen their way past revolutionary Iranian authorities.

"We'll have to keep testing their efficiency right up to the operation itself," I advised the team.

By studying the debriefs of the agents and officers passing periodically through Mehrabad Airport, we determined that the militants in control had not yet restored the level of efficiency needed to make this type of positive check a threat. In fact, one agent reported that as late as mid-December, the controls had been so unprofessional that the yellow copies of the forms hadn't been collected unless offered by the departing passenger. However, we had no way of knowing how long these lax conditions would prevail.

Besides choosing our exfiltration route, we had to fabricate a feasible cover story and provide documentation for a party of six Americans, male and female, ranging in age from twenty-seven to fifty-four. Their real professional identities raised other problems. As American Consul General, Bob Anders was a well-known figure among Iranian officials. His deputies, Joseph Stafford and Mark Lijek, were also quite familiar to a broad cross-section of Iranians who had applied for visas to America, and we suspected that our diplomats and their wives were on a Kometeh wanted list, with their pictures hidden under Mehrabad immigration counters.

Consulting with Hal, who had returned to Headquarters as the Near East Division's Chief for Iran, I immediately learned the DO's position on the cover and passport debates now under way. Even though OTS had an ample selection of foreign documents, none of the six houseguests had been trained in the fundamental tradecraft necessary to put on a convincing show at a rigorous immigration control. "We can't have them stammering their way through with some B-movie accent," Hal said. Besides, he added, many Iranians spoke foreign languages, and someone might challenge the six to respond in their "native" tongue. "They just won't be able to sustain a foreign cover story."

"Well," I quipped, trying to break the tension, "we can't exactly make them American missionaries who wandered into Iran by mistake in the middle of this crisis and now just want to go home."

Although everyone involved in OTS and the DO agreed that building a cover around U.S. passports would draw undesirable attention to the subjects, the Agency had to keep that option open. Ironically, there had been a constant stream of American journalists and well-intentioned humanitarians passing through Tehran since the embassy takeover, but we knew the Kometeh kept close tabs on those individuals. Allowing our houseguests to retain their American identity was an extremely risky proposition.

That left the Canadian option open. The six were hiding in Canadian diplomatic residences and spoke North American English, so constructing plausible cover legends around Canadian citizenship was feasible. There was one small problem, however: The government in Ottawa was constrained by laws that prohibited foreigners from using Canadian passports for any purpose. They were already challenging diplomatic convention by giving our people sanctuary.

"I don't think Ottawa is going to bend on this one," Hal cautioned.

"I'm afraid you're right," I said.

However skeptical I actually felt, I recommended the use of Canadian passports as the first choice. For the second option, I suggested using some type of foreign passport, preferably from Anglophone countries. We then plunged into a whirlwind of impromptu meetings and heated arguments, with fatigue and frustration raising stress levels. Anxiety reached an all-time high when the militants occupying the embassy discovered that two OTS-produced foreign travel documents had not been shredded when the compound was overrun. These papers had been issued to Agency officers assigned to the embassy, and one of these

officers was among the hostages. He was now subjected to brutal interrogation.

Under Khomeini's influence, radical subordinates such as the Ayatollah Mohammed Beheshti worked the mob at the embassy into a
frenzy of paranoia. They vowed to kill every hostage if America attempted to free them by force. Finding the alias documents only heightened the militants' paranoia. Foreign Minister Abolhassan Bani-Sadr
deemed the American embassy to be nothing but a "spy nest . . . a vital
spy center," from which America had secretly ruled his country through
the corrupt puppet Shah for thirty-five years.

My team and our OTS superiors seized this moment to push for
Canadian alias passports through CIA's Near East Division management. We then began an "all-sources" quest for information on the types
of individuals and groups currently using Mehrabad Airport. As a fallback, the Near East Division was developing sources for overland
"black" exfiltration options, hoping to establish contact with smugglers
whose rat lines followed safer, and less weather-dependent routes out
of Iran.

One possible source was H. Ross Perot, the Texas billionaire, who
had used a land route to exfiltrate two of his employees imprisoned early
in the Iranian Revolution. Perot had already offered his services to the
Agency to rescue all the hostages. "What's the holdup?" he'd snapped in
his usual peckish manner. "If it's red tape, I'll put up the money and you
can pay me back later."

We soon learned that the groups still traveling legally to Iran included oil field technicians from companies based in Europe, who flew
in and out of Mehrabad almost daily. Individual reporters and television
teams from all over the world covering the hostage situation also frequented the airport. Surprisingly, a number of bona fide curiosity-

seeking tourists obtained visas and traveled easily to Tehran, and the flow of self-appointed humanitarians, many of whom were U.S. citizens, continued. But all of these groups had undergone careful inspection by airport Immigration, with Kometeh agents hovering in the background. Trying to disguise the six houseguests as oil company employees or members of the news media was simply too risky. After much deliberation, I came to a decision: I would lead a small, highly experienced team directly into the revolutionary lion's den to determine if we could in fact rescue these six helpless Americans without placing them in even greater jeopardy.

"We'll have to talk to the boys upstairs," Hal said, lending me his tacit support.

After a pivotal meeting with senior Near East Division management on January 2, 1980, to present our position and review options, I leaped on a flight to Canada, accompanied by "Joe," an OTS documents specialist. We carried photos and a variety of alias bio-data for the six houseguests, so that we could show our contacts in the Canadian government how convincing their national cover would be, should they provide us with valid blank passports.

Ottawa was chill and snowbound, the Rideau Canal frozen solid, reminding us of the Moskva River in the winter. But we were welcomed warmly, and I saw immediately that our Ottawa contacts saw themselves as allies in the rescue effort. All the cable traffic about the six houseguests had passed through Ottawa to Canadian ambassador Ken Taylor, so he felt like a full participant in the operation, not just an observer. From what I could gather, he possessed many of the operational qualities we could need on the ground in Tehran: He knew how to think ahead and keep a secret.

Our first meeting that morning opened with an unexpectedly pleasant surprise. "Lon Delgado," my local liaison contact, showed me a classified memorandum, saying "Cabinet convened a rump session of Parliament to pass an order in council to approve issuing your six diplomats Canadian passports for humanitarian purposes."

I pushed our luck and requested six spare passports to ensure that we had two cover options. We then asked for two additional Canadian passports to be used by CIA "escorts." Although they approved the six redundant spares, Parliament politely declined to make an exception to their passport law to cover professional spies like me.

At our next meeting with Lon Delgado, I put forth a concept for a cover legend that had occurred to me at home when I was packing my bag for the trip to Ottawa.

"In the intelligence business," I explained, "we usually try to match cover legends closely to the actual experience of the person involved. A cover should be bland, as uninteresting as possible, so the casual observer, or the not-so-casual immigration official, doesn't probe too deeply."

I emphasized that the situation in Tehran was extraordinary, given the size of our party and their lack of experience. Therefore, why not devise a cover so exotic that no one would ever imagine a sensible spy using it?

Without citing him by name, I outlined my long involvement with Jerome Calloway, who had already volunteered his expertise to help rescue the American hostages in Tehran. In fact, the CIA had recently awarded Calloway the Intelligence Medal of Merit in a secret ceremony, making him the first nongovernment employee to be so honored. That award, he'd told me, was more precious than any of his overt professional achievements.

In my Ottawa hotel room the night before, I had called Jerome at his home in Burbank. He had no idea what I was working on and pointedly did not ask on the telephone.

"I'm in Canada," I simply told him. "I need to know how many people would normally be in an advance party scouting an overseas location for a motion picture production."

Jerome's response was immediate. "I read you . . . about eight."

"What would be their individual jobs?"

He ticked off the site scouting team: "A production manager, a cameraman, an art director, a transportation manager, a script consultant . . . that might actually be one of the screen writers . . . an associate producer, probably a business manager." He paused. "Oh, yes, the director."

He explained that the team's purpose would be to examine the shooting site from an artistic, logistical, and financial point of view. The associate producer represented the financial backers, while the business manager investigated the local banking arrangements, since even a ten-day shooting schedule could mean millions of dollars spent in local currency. The transportation manager's job was to rent a variety of vehicles, from limousines for the stars to flatbed trucks and mobile cranes for constructing the sets. The production manager was responsible for bringing all these elements together, while the other team members dealt with the actual cinematography, creating the film footage from the script.

Given America's enormous cultural influence, almost every sophisticated person in the world, including officials in prerevolutionary Iran, understood that a Hollywood production company would have to travel around the world in search of the perfect street or hillside for particular scenes. Film companies, we agreed, were often composed of an international cast and crew, and the government-subsidized motion picture and television industry in Canada had a strong international reputation.

After the meeting, I sent Headquarters a cable outlining our progress with the Canadians and presenting the movie-team option, as well as two other options: The six could pose as a group of Canadian nutritionists conducting a survey in the third world, or a group of unemployed teachers seeking jobs at international schools in the region. I was half expecting a flaming rocket in response to my suggestions, but Headquarters remained silent, always a good sign.

Over the next ten days, I shuttled between Washington and Ottawa, struggling to flesh out the complex logistical details of all three exfiltration cover options. I helped form an OTS team in Ottawa to work on the documentation and disguise items, which the Canadians had agreed to send to Tehran by courier. In Washington, my GAD team labored around the clock, collecting and analyzing the latest information on Iranian border controls. All the messages between Headquarters and the field were transmitted with the FLASH indicator, CIA's highest precedence. The sense of being engaged in a wartime effort intensified daily.

Headquarters recognized that the Hollywood scenario had potential beyond the possibility of rescuing the six diplomats. Parallel to our effort, the Agency was aiding the Pentagon in developing a military option to rescue the main body of hostages. If the movie cover held up for the exfiltration of the six, it might be possible to approach the Iranian Ministry of National Guidance with a proposal to shoot the film sequences in and around Tehran. Such a plan was not as crazy as it seemed. In spite of the embassy occupation, a number of Westerners not closely connected to the "Great Satan," America, were still doing business in

Iran. From the Iranian point of view, hosting the production of a film would have practical advantages: International sentiment was building against the Islamic Republic, while economic sanctions and the American freeze on Iran's assets in the U.S. was starting to inflict damage. Allowing a Canadian production company to shoot a film would create a facade of normalcy while encouraging a flow of hard currency.

Between trips to Ottawa and intense planning sessions with the Near East Division, I made a quick trip to consult Jerome Calloway in California. I brought $10,000 in cash with me, the first of several "black bag" money deliveries to set up our cover motion picture company. Normally, such expenditures would be considered wasteful, since the Carter administration had not yet approved the movie option. But Admiral Turner and the Deputy Director for Operations realized that we had to be flexible, and that preparing such an elaborate facade took time and money.

Jerome met me at the airport on Friday night and introduced me to one of his Hollywood associates, "Robert Sidell," at a suite of production offices they had managed to claim for our purposes on the old Columbia studio lot in Hollywood. With intensive effort over the next four days, Jerome and his associates managed to create "Studio Six Productions" by the close of business on Tuesday. Our offices had just been vacated by Michael Douglas's *China Syndrome* production company, and we were sitting on prime movie country real estate. Jerome's team included masters of the Hollywood system, and they had begun distributing what they called "grease" (cash payments and calling in favors), even before I arrived. Simple things such as the installation of telephones and furniture rental often took weeks and involved the industry's turf-conscious web of unions. But we had everything we needed to operate by Tuesday afternoon.

On the second day of Studio Six's conception, we turned to the important question of identifying an appropriate script. Jerome and I were sitting at his kitchen table discussing possible themes when he suddenly looked up beaming. "I think I've got what we need," he said. He had recently received a script a little too evocative of *The Exorcist* for his taste, but which combined mystical elements with nonspecific Middle Eastern locations.

Jerome shuffled through the heap of screenplays littering his office and retrieved the script in question. It fit our purposes beautifully, particularly since its complex stage directions and cinematography jargon would be almost unintelligible to the lay reader. The script was an adaptation of an award-winning sci-fi novel and had already attracted some interest in the industry. Unfortunately, the producers, who had planned to transform the film's massive set into a major theme park, had run into financial problems, and the project had collapsed. But they had hired a famous comic-strip artist to create elaborate visual storyboards of the sets and scenes, giving us more eyewash to bolster the production portfolio.

Now we had to find the appropriate name for our "property." I had already decided that my cover in Iran would be as the production manager, and I would carry the only full copy of the script to use as a prop to show the Iranian authorities if necessary during our exit through Mehrabad Airport. But we still had to repackage the script with an appropriate logo and title.

Over more coffee, the three of us searched for the perfect name for our bogus movie. We needed something catchy, something evocative of the Middle East or mythology. After several failed attempts, Jerome came up with the winner, resorting to the punch line of our favorite knock-knock joke.

"Let's call it *Argo*," he said with a wry grin.

"Argo"—a contraction of "Ah, go fuck yourself," mumbled by a drunken bum in the joke—had often been used to break the tension of working long hours under difficult circumstances. But Calloway also noted that Argo had major mythological connotations: Jason and his Argonauts had sailed aboard the vessel *Argo* to recover the Golden Fleece, after performing many feats of heroism.

"Sounds pretty much like this operation," I said optimistically.

I quickly designed an *Argo* logo on a yellow legal pad, and the next day, we ordered full-page ads announcing the upcoming Studio Six production of *Argo* in *Variety* and *The Hollywood Reporter*, the two most important "trades" of any show biz publicity campaign. The Studio Six production of *Argo* was described as a "cosmic conflagration," from the story by Teresa Harris—the alias we had selected for our story consultant, who would be one of the six houseguests awaiting rescue in Tehran.

On my last day in California, I called the Iranian consulate in San Francisco, using my operational alias "Kevin Costa Harkins." The officials at the consulate, like most Iranian diplomats in America, were vestiges of the former Shah's government and existed in a state of limbo.

"I'm going to need a visa and instructions on how to obtain permission for scouting filming locations in the Tehran bazaar," I told the befuddled Iranian consular officer. I added that my party of eight would be made up of six Canadians, a European (myself), and our associate producer, a Latin American.

"Oh, sir," the man said meekly. "There is nothing we can do to help you or your other gentleman. The situation here is not normal. I suggest you apply to another Iranian consulate."

Here was a minor complication. "Julio," the OTS authentication officer who would accompany me, was already posted overseas, where he could receive his visa. He had Arabic language skills and considerable exfiltration experience. He also spoke excellent German and fluent Spanish, and could carry foreign passports with either ethnicity. We had already documented him with a cover legend as one of our production company's South American backers. Since I planned to travel on an alias European passport, I would just have to obtain my Iranian visa abroad.

Flying back to Washington on the red-eye that night, I carried a rich collection of Hollywood pocket litter, right down to matchbooks from the Brown Derby, where Studio Six Productions had hosted a bon voyage dinner for me. At the airport, Jerome threw his huge arms around me, muttering, "Take care of yourself over there, Tony." It was an unusually affectionate gesture for the gruff Hollywood veteran, whose style tended to be more evocative of a night club bouncer than a makeup artist.

AT THE CONCLUSION of a series of meetings back in Washington and in Ottawa, Lon took me aside and explained in a friendly but concerned manner that the Canadians were losing patience with Washington. "Look, your government hasn't even stated a preference on your operational plan," he said. "But Ottawa has made a series of unique concessions without hesitation. What's taking you so long to move down there?" He added that the six houseguests, who had been hiding in the Sheardowns' villa and Ambassador Taylor's residence, were becoming increasingly nervous. Unusually large bands of Revolutionary Guards had prowled around the Shemiran neighborhood, and a military helicopter had been sighted flying low over the Sheardowns' house, terrifying the Americans and Zena Sheardown.

The most disturbing turn of events involved possible press leaks.

Earlier in the crisis, Jean Pelletier, the Washington correspondent of Montreal's *La Presse*, had surmised from the official State Department hostage list that a number of American diplomats were unaccounted for in Tehran. Pelletier contacted the Canadian ambassador in Washington, a family friend, who confirmed his suspicions, but prevailed on the young journalist to sit on the explosive information until after the exfiltration. In return, the Canadian government promised Pelletier an exclusive on the dramatic story.

But other journalists were chewing around the edges of the secret, trying to unravel the purposely vague information on hostages that our State Department spokesman had provided as a smoke screen. It was clear that the news media could not be held at bay indefinitely.

Then, one afternoon in Tehran, Ambassador Taylor's wife, Pat, answered a call on the personal line in the residence.

"Could I please speak to Joe or Kathy Stafford?" The man was soft-spoken, with a North American accent.

Pat Taylor had stifled momentary panic and calmly replied, "Is this some kind of a joke? There's no one by that name here."

"Look," the man persisted, "I know the Staffords are staying at your home."

It was impossible to determine whether or not the call had originated from inside Tehran or overseas. But one thing was certain: The net was drawing in tighter. The Canadians saw their situation becoming tenuous and initiated plans to close down their embassy before it, too, was overrun and their diplomats taken hostage.

■

NOW JULIO AND I shifted our attention to Europe. Agency officers there had been actively debriefing travelers and collecting the most up-to-date intelligence on Iranian document controls. As this process took place, Julio and I worked in separate cities, each preparing his own alias document package.

We planned to join up in Frankfurt for our final meeting before our launch into Tehran, now tentatively set for January 23 and 24, 1980. Julio and I would apply for Iranian visas at different embassies. If neither one of us was successful, I had a fallback position: A CIA officer in the region had an alias passport on which he had received a visa from a large Iranian embassy staffed by the new revolutionary diplomats. I could borrow his identity and alias passport if I had to.

On Monday, January 21, Julio left Frankfurt for Geneva to apply for an Iranian visa. I departed Washington's Dulles airport the same day, traveling under my real name on my U.S. official passport but hand carrying the Studio Six portfolio and the materials I would need to complete our document packages.

I met Julio in Frankfurt on the morning of January 22. He brandished his alias passport with its newly stamped Iranian visa. "No problem, Tony," he said happily. "They seemed happy to have me visit their country."

Now I had to obtain my own visa. I planned to drive to Bonn the next morning, and hoped the Iranians there would also be accommodating.

That afternoon, we received a FLASH message from Ottawa. Our exfil kits had arrived in Tehran, but Ambassador Taylor and his first secretary, Roger Lucy, had reviewed the documents and discovered a serious mistake. The handwritten Farsi information on the rubber-stamped Iranian visas produced in Ottawa showed a date of issue in February, three weeks in the future. Our Farsi linguist had apparently misconstrued the

Shiite Persian calendar. This type of error could unhinge the most meticulously planned operation, but it could also be easily corrected.

We fired back a message through Ottawa, assuring Taylor that we would deal with the mistake once we arrived in Tehran.

On Wednesday morning, January 23, an Agency officer drove me to the Iranian embassy in Bonn. For the first time, I was struck with a mixture of exhaustion, jet lag, and fear. Having been in Tehran for the RAPTOR exfiltration only eight months earlier, I worried that a paper trail of my presence there might have surfaced among the militants occupying the embassy. We knew that they had assigned large teams of child carpet weavers to the painstaking task of reassembling the shredded documents, just so they could identify American intelligence officers who had operated from the building.

I had altered my appearance slightly to match my alias passport photo and brought along the Studio Six portfolio as a prop. To complement this disguise, I wore a green turtleneck under a well-cut European corduroy sportscoat, hoping to evoke an artistic, unmenacing persona.

As we approached the Iranian embassy, I was alarmed to see that the embassy of my supposed country of origin was just down the street. If the Iranians doubted my cover, it would be perfectly standard for them to send me to my "own" embassy for a letter of introduction before granting the visa—something I dreaded, since my passport was spurious. I'd have to resort to the fallback alias passport, which was a prospect riskier than traveling on a freshly minted, authentic visa.

I left the car a block from the Iranian embassy and tried to stroll casually to the consular section. There were about a dozen visa applicants sitting in the reception room completing forms. Grim-faced young Revolutionary Guards in ill-fitting civilian clothes stood in the corners, glaring at the heathens like myself. I took a chair and began to fill out the application, then felt myself become flushed with anger and fear. I had left the Studio Six portfolio in the car when I was dropped off. All I had to convince a stern Iranian security officer that I actually was a movie producer was my alias passport and a few other personal identity documents, including my Studio Six business card. Taking a deep breath, I carried the forms to the clerk's window and presented them to the consular official.

"What is the purpose of your visit?" he asked with quiet deference.

"A business meeting with my company associates at the Sheraton Hotel in Tehran," I said in my most polished accent. "They are flying in from Hong Kong tomorrow and are expecting me."

He studied my application and flipped through my passport, noting the tangle of visas and entry and exit cachets. "Why didn't you obtain a visa in your own country?"

"Because I was here in Germany on business when I received the telex about the Tehran meeting." I shrugged casually. "I have to fly to Tehran tonight."

He nodded. "Very good."

Fifteen minutes later, I left the Iranian embassy with a one-month visa stamped into my alias passport.

The operations plan for my entry into Tehran called for me to board a Swissair flight from Zurich, scheduled to arrive at five A.M. on the twenty-fourth. Julio would follow exactly the same itinerary one day later, so if anything happened to one of us en route, the other might still get through.

Back in Frankfurt, I sent a FLASH cable to Washington and Ottawa that I was ready to depart. Within minutes, I received approval to launch that afternoon. But thirty minutes later, a communicator handed me yet another FLASH, this one from DIRECTOR, asking me to suspend my departure because President Carter wanted to be briefed once more on the entire operation before granting final approval to the covert operation.

I paced the windowless office, struggling to overcome my mounting tension. Carter was a notorious stickler for detail. Would he cancel this entire operation—because he didn't understand some arcane point of our profession?

Half an hour passed before the communicator reappeared with a final message from DCI Turner: PRESIDENT HAS JUST APPROVED THE FINDING. YOU MAY PROCEED ON YOUR MISSION TO TEHRAN. GOOD LUCK.

While I was packing my bag that afternoon at the hotel, someone rapped softly on my door. Julio ushered in an older man, a near-legendary Agency contract officer who had been traveling in and out of Tehran in support of the larger hostage rescue operation. Working as a "NOC" (Non-official Cover), his main responsibility was to create the internal support structure for the Delta Force, a highly risky endeavor.

The man's usually long white hair had been clipped and dyed for the mission, giving him the appearance of robust middle age. Unless you knew the reality, it would be hard to guess he had been in the covert action business since World War II, when he had parachuted into the Nazis' Fortress Europe to work with resistance groups. Having just returned from Tehran, he gave us very useful insight into the latest situation at Mehrabad and in the city. His most important information concerned documentation.

"There're still a lot of Iranians trying to get out with forged papers," he explained. The fact that we would be using valid passports, some with valid visas and others with the best possible forged replicas, was reassuring. But more than this hard intelligence, his positive spirit bolstered our confidence. This man was a master at our craft, and he had just thrown holy water on our hazardous endeavor.

My Lufthansa flight from Frankfurt departed exactly on time, and I allowed myself the luxury of a scotch. The Swissair flight from Zurich to Tehran was normally never delayed, so as I sipped the scotch in the Lufthansa Boeing, watching the moonlight flicker on the snowy Alps below, I tried to calm my inner turmoil. I was committed. The operation was perfectly planned. In two hours, I'd be en route to Mehrabad.

But once in the Zurich airport terminal, waiting for my Tehran flight to be called, I found myself pacing among the passengers. The flutter of nerves I had battled flying down from Frankfurt returned. Every minute that passed before boarding the plane gave me time to consider the gravity of the situation, and I was suddenly overcome with cold panic.

Although I could share only certain details of past exfiltrations with Karen, she had watched me become increasingly anguished and exhausted as this operation unfolded. And when I left the States, neither of us had spoken much about my assignment, although she knew that I was going into Tehran under alias. There had been no need for such a conversation. Karen understood the nature of my profession, and she also realized I would never send a less experienced officer to Tehran if I believed I had the crucial expertise on which success might depend. But on my last morning in our mountain valley, we were both overcome by an unspoken sadness. Again, events had conspired to take me away from my family. As was the usual practice when leaving on this kind of operation, the last thing I did when she dropped me at the airport was to hand her my wedding ring. Agency spies traveling under alias always used legends of single people.

It was this heavy sadness, rather than fear, which paralyzed me now.

For several minutes, I trembled, cringing inside my overcoat. Then, the emotion passed.

Suddenly, the loudspeaker crackled. My flight to Tehran had been canceled because of a snowstorm that had closed Mehrabad Airport.

Chuckling at the irony, I checked into a hotel, ordered a steak with a bottle of wine, and called Julio in Germany. We would stick to his schedule the next night, arriving in Tehran together.

"I guess Murphy is still riding with us," Julio joked.

"Roger that."

WE ARRIVED AT Mehrabad at five A.M. on January 25. The DC-8 taxied between heaps of freshly plowed snow. As we walked to the bus, the coal smoke and lingering smog of winter in Tehran swept across the tarmac, an eerie reminder of the subcontinent's smit.

Much to our surprise, immigration and customs controls were quick and efficient. The officer to whom I handed the disembarkation/embarkation form with my passport was a uniformed professional, not a disheveled, untrained irregular of the type I had encountered in April 1979. The man hardly glanced at our passports before slamming down his cachet entry stamp and tearing apart the white and yellow sheets of the forms.

I noticed some plainclothes Kometeh security types in the terminal, but they ignored the foreign passengers and concentrated on the Iranian families returning from abroad. I knew that the Kometeh and Revolutionary Guard presence would increase in an hour or so, as passengers leaving on the return Swissair flight began to appear. If the stream of families seeking exile continued, the Revolutionary authorities would be watching for people smuggling gold or valuable carpets—the ubiquitous emergency currency of the Middle East—out of Iran.

Our cab driver, the proud owner of a vintage Opel, was very friendly. "Eat food," he shouted, handing us a piece of unleavened bread piled high with crumbly goat cheese.

As we careened on the roads to the Sheraton Hotel, we noted Farsi propaganda banners and revolutionary posters draped on the buildings along the main avenues. Fortunately, checking into the hotel at this early hour posed no problems, and a quick scan of the lobby revealed no overt security, although in the back of my mind I sensed that the Kometeh was present.

The next step on our ops plan was the nearby Swissair office, where I planned to reconfirm our eight reservations on the Monday morning flight to Zurich—a vital part of any exfiltration. But the Swissair office was not yet open. Since I knew the U.S. embassy was right down the street, I suggested we stroll past it, then reconnoiter the Canadian embassy, which our tourist map indicated was nearby.

It was bizarre to walk up the broad avenue toward the familiar brick-walled compound, now plastered with more propaganda photos of the Ayatollah Khomeini with his white beard, glowering beneath his black turban. The wrought-iron gates were studded with loudspeaker horns, which called the faithful to daily prayer and led them in the familiar chant: "Death to the Shah! Death to America! Death to the CIA!"

As we approached one of the sandbagged guard posts beside the main gate, I felt another tremor erupt along my spine as reality hit me. Fifty-three of my fellow citizens were being held in that building or at the nearby Foreign Ministry. Although a small group of women and blacks out of the original sixty-six hostages had been released before Christmas, the situation was still grievous. Some of the hostages in the embassy were CIA officers, and I knew the militants reassembling the shredded documents would eventually discover their real identities and turn on my colleagues with a vengeance. At this point, there was nothing Julio or I could do but memorize the exact position of the sandbagged machine-gun posts and observe that the roofs of the compound buildings were free of antiaircraft weapons.

We continued down Roosevelt Avenue and turned left onto Motahari, which the Revolutionary Council had apparently renamed for one of their martyrs. Julio and I stopped at the curb, our tourist map open between us, searching for the Canadian embassy. There was a building with a flag across the street, but it flew the blue and gold of Sweden, not the Canadian red maple leaf.

The uniformed policeman at the entrance of the Swedish embassy could not understand our questions. Julio tried German, Arabic, and even Spanish, but nothing worked. We handed the policeman our map, and I slowly pronounced the syllables, "Can . . . ah . . . dah." But the guard's perplexed frown only deepened.

The CIA had some of the best mapmakers in the world, as did the Defense Department. But our Studio Six cover was sure to be blown if the Revolutionary Guards searched us and found an intricate U.S. government–produced map. Operational security had required that we use a tourist map purchased at an airport kiosk in Europe. As I always told younger officers, you had to live your legend.

At that moment, a young Iranian crossed the street to join the discussion. He was bearded and wore a faded army field jacket over rumpled civilian clothes. To me, the young man looked suspiciously like one of the "students" who was occupying the embassy. He spoke sharply to the guard, and seemed to be demanding an explanation for our presence. Then he turned to Julio and addressed him politely in German. They opened the map together and immediately fell into a friendly discussion, the young man speaking in fluent German.

Is he just pinning us down here until his buddies show up? I wondered. But the man seemed perfectly polite and helpful. He borrowed a
pen and a page from my notebook and jotted down an address in Farsi.
Then he motioned for us to follow him back to Roosevelt Avenue, where
he flagged down an old Mercedes taxi. After speaking intently to the
driver in Farsi and showing him the slip of paper, the young man held
open the door for us to get in.

"Danke schön," Julio said, offering the young man some crumpled
rial bills.

The fellow refused graciously, placing his hand on his heart and
offering a gold-toothed smile, as if to indicate that the Revolution, faithful to the true tenets of Islam, was grounded in hospitality. It was hard
to reconcile this image with the brutal reality. Less than a block away,
Americans were locked in closets or tied to chairs, floundering in a black
hole of terror and uncertainty.

The taxi seemed to travel half the city before dropping us off at a
new building marked with the seal of the Canadian embassy. Ambassador Ken Taylor had been expecting us, and he was waiting in his outer
office when a husky military police sergeant named Claude Gauthier
escorted us upstairs. A lean man in his forties, wearing jeans, a plaid
Western shirt, and gleaming cowboy boots, Taylor hardly fit the stodgy
image of an ambassador. Most incongruous of all were his mod Italian
glasses and helmet of tight salt-and-pepper curls. Reaching out to shake
our hands, Taylor smiled benevolently, baring perfect white teeth.

"Welcome to Tehran, gentlemen," he said with natural charm.

I saw that he was completely at ease and hardly the tense bureaucrat
I had imagined. Ken graciously introduced us to his secretary, a petite
elderly woman named Laverna, who offered us a cheerful, relaxed greeting. We could have been two technicians hired to fix the air conditioner,

THE MASTER OF DISGUISE 293

not American spies assigned to rescue six hidden diplomats, for all the casual friendliness of this encounter.

In Ken's inner office, his manner became more serious. "I've pared down my staff almost completely," he explained. "They left Tehran in small parties. Once my family flies out this afternoon, there'll be only five of us left. But we're scheduled to depart Monday morning on British Airways to London, just after your Swissair flight with the houseguests on board."

Taylor said that he planned to send the Iranian Foreign Ministry a diplomatic note by messenger early Monday morning, explaining that the Canadian embassy would be "temporarily" closed. "Now," he asked, "what can I do to help you fellows?"

Reciting from the operations plan I had memorized to the last detail, I laid out the tasks we would need to accomplish in the next three days. "Our first order of business is to meet with the houseguests, brief them on the plan, and assess their ability to pull it off."

"Let's meet tonight at the Sheardown villa," Taylor suggested. "My counselor, Roger Lucy, can drive the Staffords there from the residence. So far, the Revolutionary Guards haven't harassed our diplomatic vehicles, so I think that's safe."

Before we left the embassy, Taylor informed us that two "friendly" ambassadors and their staffs had also become involved in harboring the six Americans. These ambassadors had joined with Taylor in his regular visits to the American chargé, Bruce Laingen, and two fellow Foreign Service officers, who were under "protection" at the Iranian Foreign Ministry. In theory, these diplomats, who had gone to the ministry to protest the demonstrations before the takeover, were free to leave Iran at any time, but they refused to abandon their colleagues held in the compound.

Taylor sent a message to Headquarters via Ottawa, confirming our safe arrival and outlining our plans. Before leaving the embassy, Taylor introduced us to his number two, Roger Lucy, a confident young man who had been caring for the four houseguests at the Sheardowns' residence after the couple had departed earlier that week.

"I see you've met Sergeant 'Sledge,' " Lucy quipped.

The burly Quebecois MP actually blushed. He'd earned that epithet wielding a twelve-pound sledgehammer, destroying all but the most essential classified cryptographic and communications equipment in preparation for the embassy shutdown. Sergeant Gauthier had also been on an informal recon near the American embassy a few days earlier when militants had dragged him inside for interrogation. He had quickly grown tired of the rough treatment, stood up, knocked them aside, and stalked out, leaving them in shocked silence.

It was Sergeant Sledge who drove Julio and me to the Sheardown villa in Shemiran that evening. Although it was dark, I could tell that the house was spacious and securely protected from the street by a high stucco wall.

As soon as Roger Lucy opened the door, the six rushed forward to meet us, beaming with anticipation, as if Julio and I alone carried the keys to their freedom. Once more, I felt as if I had stumbled into the Twilight Zone. A fire burned merrily on the hearth, and there were trays of hors d'oeuvres on the inlaid cocktail tables. As Roger Lucy poured drinks, I observed that the six showed no obvious signs of severe anxiety.

One of the couples, Joseph and Kathleen Stafford, sat on a couch opposite Mark and Cora Lijek, while the two single men, Bob Anders and Lee Schatz, stood near the fire. Anders, the consul general, looked fit and had even retained a tan. He explained that he had been able to exercise in the Sheardowns' walled garden almost daily until the helicopter incident the week before. Lee Schatz, the agricultural attaché, a tall, younger man with a bushy mustache, appeared to be someone with quiet strength, a potential leader.

Once we had gotten acquainted with each other, I rose to brief the six on the preparations we had made for their escape.

"We have three separate cover stories," I explained, "each with their own passports and supporting documents. You will have the final choice, but Julio and I can certainly advise you."

The six began to pepper us with questions. How did we know what the airport controls would be? How would they answer if taken aside and interrogated about their presence in Tehran? I patiently answered each of these questions, and almost everyone appeared satisfied. But Joseph Stafford still seemed uneasy about the risks involved. Highly intelligent, he failed to exhibit the spirit of adventure kindling among the others.

Mark Lijek pointed to the U.S. passports and itinerant-teacher documents stacked beside the other two option packages on the table. "For openers," he said, "I think this teacher deal is just plain crazy." The others nodded vigorously, and I slid those documents to one side.

"I've managed a lot of these operations," I said with my most confident smile, "and I believe the *Argo* movie plan will work."

I opened the Studio Six portfolio and spread out the full-page ads from the Los Angeles trade papers. Then I handed Cora Lijek her business card and pointed to the page from *Variety:* " 'Based on a story by Teresa Harris,' " I recited. "You've just become Teresa Harris." I flipped open her *Argo* cover-legend Canadian passport to the identity page. She stared at her picture with its forged signature.

"If anybody calls Studio Six Productions in Los Angeles," I added, "they'll be told Teresa Harris is with the location scouting team in the Middle East but will be back next week."

I prayed that I was beginning to win them over.

Her husband, Mark, leaned over Cora's shoulder. "Well," he said, "the movie crew idea isn't *totally* crazy."

"Look," I finally suggested, "get together in the dining room and discuss these options among yourselves. First, try to decide if you want to leave as a group or individually because we've only got tomorrow morning to reshuffle your plane reservations."

I waited about fifteen minutes, then entered the dining room. As I had expected, Joe Stafford was not impressed by any of the options. After a few minutes of more debate, Bob Anders called for a vote, then turned to me. "We've decided to leave together as a group with the Studio Six cover."

We were able to relax for the time being, so Lee Schatz and Cora Lijek showed us around the villa, which was vast and beautifully furnished, with a kitchen usually found in luxurious restaurants. Patting his stomach, Lee said that they had spent much of the previous twelve weeks planning and cooking gourmet dinners for themselves and the few trusted outsiders who visited.

"We've also become the Scrabble champions of Iran," Cora said cheerfully.

At that moment, one of the friendly ambassadors and his attaché, "Richard," arrived to meet us, confirming their intention to help in any way they could, news we received with gratitude. Before Claude drove us back to the city center, I issued the Six their *Argo* supporting documents and résumés from the portfolio, wishing them luck and feeling optimistic as I looked into their hopeful faces.

■

OVER THE NEXT two days, the operation progressed swiftly. While the six were rehearsing their legends, Julio and I worked at the Canadian embassy on the travel itinerary of the Studio Six team, whose cover legend and airline tickets showed them arriving in Tehran from Hong Kong within the same hour that Julio and I had arrived at Mehrabad from Zurich. Because this Air France flight had landed on time, its disembarking passengers would have been processed by the same immigration officers we had encountered. Therefore, the Iranian cachets stamped in our passports were prime exemplars for those we now entered in the passports of the six.

That afternoon, we asked Richard to drive out to Mehrabad and pick up a stack of dual-sheet immigration forms from one of his contacts working for an international airline. Julio completed the Farsi notations on about twenty of these, using phrasing identical to our own. Such an act provided the six with enough forms to enter their false bio-data in their own handwriting and to use their alias signatures.

On Sunday morning, I completed a long cable to the Agency describing the final operations plan in detail, including Taylor's proposed addition to our cover:

SIX CANADIANS FROM STUDIO SIX PRODUCTIONS CALLED ON THE AMBASSADOR, HOPING HE COULD FACILITATE AN APPOINTMENT WITH THE MINISTRY OF NATIONAL GUIDANCE TO PRESENT THEIR PROPOSAL TO LEASE PART OF THE LOCAL BAZAAR FOR TEN DAYS DURING THE SHOOTING OF THE FILM *ARGO*. THE AMBASSADOR HAS ADVISED THEM TO SEEK A LOCATION ELSEWHERE IF POSSIBLE, BUT HAS OFFERED ONE OF THE EMBASSY'S VACANT RESIDENCES AS GUEST QUARTERS UNTIL THEY CAN ARRANGE TRAVEL.

THEY HAVE ACCEPTED THE EMBASSY'S HOSPITALITY AND ADVICE AND WILL
PROBABLY DEPART ON MONDAY, 28 JANUARY.

When we reconvened Sunday night at the Sheardown residence,
Julio and I were struck by the transformation of the six's appearances
and personalities. On Friday night, I had given them the disguise materials and clothing props we had been able to secure without arousing
suspicion. Since then, they had borrowed clothes from each other and
revamped their images, having fun as they hammed it up as glamorous
Hollywood people.

Cora Lijek had parted her dark hair so that it fell back across her
neck in a severe "literary" style. Normally a nonsmoker, she now puffed
nonchalantly on a cigarette, looking the epitome of the sophisticated
screen writer. I darkened Mark Lijek's wispy blond beard with mascara,
altering a face that had become too familiar at the U.S. Consulate. Then
I showed Kathleen Stafford how to pin up her long brunette hair. After
adding thick-rimmed glasses and the *Argo* location sketchbook to her
ensemble, she could easily pass muster as the set designer.

But Bob Anders, the stereotypically conservative consul general,
had pulled off the most dramatic metamorphosis. His white hair was
now puffed out in a blow-dried pompadour, and he wore tight, pock-
etless twill jeans with a slight flare, an even tighter blue silk shirt open
at the chest, and a chunky gold chain and medallion.

"Check this out," he said with a wry smile. Slipping a topcoat across
his shoulders with the suave bravado of a character in a Fellini film, he
strutted around the room with the chutzpah of a Wilshire Boulevard
stud.

After the fashion show, we briefed them on the details of their supposed journey and arrival in Tehran. They were clearly intrigued with

the visas and the deceptively authentic Farsi notations Julio had made. I then collected the white originals of their immigration forms and burned them in the fireplace. Several of the six were concerned that exit control might try to match their yellow copies with the white cover sheets. "The authorities haven't checked the white originals for months," I assured the six. "But if they do complain, just act dumb. How are you supposed to know what happened to their little white sheets? You've got your yellow copies."

Roger Lucy tried to lighten the mood. "Just remember to end every sentence with 'eh?' and you'll be all right."

Before dinner, I warned the houseguests not to drink too much, reminding them that they would face a "hostile interrogation" after the meal from Roger Lucy. Beaming as usual, Taylor arrived soon after. He had received an answer to our cable. The policymakers in both Ottawa and Washington were pleased with our proposed plan of action. They had concluded their message with "SEE YOU LATER, EXFILTRATOR."

The two friendly ambassadors now joined us, and the mood became quite festive. The six served us a seven-course dinner, along with the Sheardowns' vintage wine, champagne, coffee, and liqueurs. When I revealed the origin of Argo and Jerome's knock-knock joke, they all raised their glasses and rallied to the "Argo!" battle cry. Then I got serious, and asked them to resist the temptation to publish details of the exfiltration in the future, as it was crucial for Julio and me to stay in business.

Now it was time for the interrogations. Roger Lucy, dressed in camouflage battledress jacket and military boots, took each American into a side room to be grilled. We heard him shouting angrily, relying on the old interrogator's trick of using the agent's real name, not the alias. Each time one of the six slipped up, Lucy began the fierce interrogation again. "Where visa issued?" he shouted in his best Farsi accent. "When? Name of father? You liar! You American spy!"

Before Julio and I left at midnight, we reviewed the final arrangements for travel to the airport. I would precede the others by thirty minutes, driven by Richard, who would pick me up at the hotel at exactly three A.M. Once in the terminal, I would confirm that the security situation was normal and that the Swissair plane was en route from Zurich. Then I would put my suitcase through customs departure control and check in at the airline counter, where I would linger so that the others could see me as a personal, "all clear" signal as they entered the terminal. Julio would accompany the six to the airport in the embassy van and lead the way through Customs.

As Julio and I shook hands all around, the six repeated their hearty "Argo!" cry. *They're as ready as they'll ever be,* I thought.

I WOKE WITH a start in my hotel room to the shrill rings of the telephone. Richard was calling from the lobby. It was three in the morning, and I should have been up at 2:15. My watch alarm had gone off, but I had slept through it. I jumped into the shower, dressed, and joined Richard in the lobby less than fifteen minutes later. He drove his ambassador's Mercedes dangerously fast through the narrow streets, and we arrived at the floodlit Mehrabad terminal just after 4:30 A.M.

The customs clerk glanced at my passport, chalked my suitcase, and nodded toward the ticket counters. There were just a handful of passengers in the hall, and several airport employees were dozing at their desks. The friendly young Swissair rep confirmed the flight was on time to arrive at five A.M. He took my bag and issued a boarding pass and baggage check. *So far, so good,* I thought.

Leafing through a copy of a magazine, I stood at my designated spot, waiting for the rest of the party. Richard went off to find his airline contact, just in case we needed a fallback flight. At two minutes after five, Julio led the six up to the customs departure counter. To my practiced eye, they looked slightly hungover and nervous. But I noticed that they had all done an excellent job applying their minimal facial and hair disguises and their stylish clothing. Once they heaved their bags, plastered with maple leaf stickers, onto the customs counter, they fell naturally into relaxed banter, a ploy that would have pleased the most critical tradecraft instructor at the Farm.

Check-in at the Swissair counter went smoothly, and now we were ready to proceed as a group to the immigration checkpoint, which was often haunted by Revolutionary Guards later in the day. If the operation was going to unravel, it would probably be here, either from hypervigilance on the Iranian side or obvious nervous behavior among the Americans.

I stayed behind, as the production manager responsible for the well-being of his team. I was armed with the leatherbound *Argo* portfolio and prepared to overwhelm any inquisitive official with Hollywood jargon.

But Lee Schatz was so eager that he had moved ahead and was already speaking with the Immigration officer. For a moment, my mouth went dry as the man studied Schatz's passport, then looked up to ask, "Is this your picture?"

"Yes . . . of course," Schatz said, his voice quavering.

The officer disappeared through a door without speaking. *Is he looking for the white disembarkation sheet?* I worried. Then the man reappeared and thrust Schatz's open passport toward him. "The picture looks different," he remarked.

I studied the demeanor of the other Americans in the stalled line, but was glad to see that no one seemed on the verge of panic. Schatz saved the day by smiling and making a clipping motion with his fingers at the ends of his mustache. "It's shorter now," he said.

The officer shrugged, stamped the passport, and handed it back across the counter. As the others lined up to present their passports and yellow embarkation forms, the officer's arm rose and fell robotically as he stamped each passport. He also collected the yellow forms but treated them so carelessly that one of the sheets floated to the floor. When no one was looking, I snatched it up. It was Anders's form, and I slipped it into my sportscoat as a souvenir.

We were soon in the departure lounge with only the final security check to pass through before boarding the buses to the plane. I was pleased that the six followed instructions and were strolling through the duty-free shops like ordinary tourists, seemingly oblivious to the sudden appearance of four Revolutionary Guards in rumpled fatigues, who broke into pairs and began to prowl the hall, scrutinizing the passengers.

But when I looked back, I was stunned to see Joe Stafford holding a Farsi newspaper. He had grabbed the first prop he could find to hide his face, but he had chosen the wrong one. Fortunately, he seemed to sense the problem and snatched up an English magazine instead.

The Revolutionary Guards concentrated on the Iranian passengers, gruffly demanding to inspect their papers. *Let's hope they're trying to flush out gold smugglers or maybe fishing for a bribe,* I thought.

Richard appeared with his airline friend. Speaking in a stage whisper, the man asked why we hadn't booked on his airline, for he would have given us the "royal treatment."

"I'll keep that in mind," I said softly. "We may need you if Swissair gets hung up."

Then the loudspeaker announced the departure of our flight. Releasing a deep sigh, I shepherded the party of Americans into the glassed-in security room at our gate. The only thing standing in the way of our freedom now was a short bus ride to the aircraft.

The PA system echoed again. "The departure of Swissair flight 363 is delayed due to a mechanical problem."

"Everything will be okay," I said trying to reassure the party as we filed back into the departure lounge. The room was filling up with passengers from flights arriving en route to Europe and Asia. How long could we press our luck here before some perceptive Kometeh counterintelligence type made his daily rounds of the departure lounge, searching for foreign spies? I thought that it was time to switch flights, because Murphy was definitely riding with us every minute of this trip.

I found Richard and his airline man in the corner of the lounge. They had already spoken to Swissair and assured me that the mechanical problem was minor, just a faulty airspeed indicator that would be replaced within an hour. We hurriedly discussed the option of switching to KLM or British Airways, but decided that would only draw undue attention to our party.

"It's best to be patient," Richard said earnestly.

Back at the benches where the others were sitting, I tried to convince them that we wouldn't have long to wait. But I could see that my subjects were on the verge of becoming unhinged. The Revolutionary Guards had switched from harassing Iranian passengers to badgering foreign travelers, questioning them rudely in broken English or German.

One of the most agonizing hours of my life dragged by as bleak winter daylight filtered through the windows. Planes landed and took off. We sat in the stuffy, overcrowded departure lounge. Finally, the loudspeaker crackled: "Swissair flight 363 ready for immediate departure, gate four."

As we clambered aboard the bus, I looked at my fellow Americans' ashen faces. What had begun almost as a lark had become mental torture.

I was still feeling exhausted, trembling from an adrenaline overdose as I wearily climbed the boarding stairs to the plane. Then Bob Anders punched me lightly on the arm. "You guys arrange everything, don't you?" He pointed to the name of this DC-8, lettered neatly across the nose: ARGAU, a city and region of Switzerland, and the perfect vessel to carry our Argonauts to freedom.

The flight crew could not serve alcohol until we had cleared Iranian airspace, so by the time we could hoist our Bloody Marys, we knew we had successfully escaped the brutal reach of the Kometeh.

IN ZURICH, SEVERAL of the six dropped down and kissed the tarmac at the base of the *Argau*'s boarding stairs—an act which raised eyebrows among the staid Swiss. Once past Zurich formalities, State Department officers whisked the six away in a van to a debriefing lodge in the surrounding mountains. They left Julio and me standing in the parking lot.

I was still in Frankfurt working on my postaction report when Jean Pelletier's *La Presse* story broke in Montreal. The news spread fast to the world media. Two days later, when I arrived at JFK, I bought a copy of the *New York Post* bearing the triple-decker headline CANADA TO THE RESCUE!

In the weeks that followed, the United States, frustrated and enraged by the hostage stalemate, unleashed a flood of gratitude on Canada. Maple leaf flags and billboards emblazoned with THANK YOU CANADA shot up across the country, and Ken Taylor became an instant hero and celebrity nicknamed "the Scarlet Pimpernel" of diplomacy.

Aside from a few vague media references to unspecified assistance from the CIA, the Agency received no public credit for conceiving and directing the operation.

◾

ON MARCH 12, 1980, I accompanied Admiral Stansfield Turner, the Director of Central Intelligence, to his morning meeting with President Jimmy Carter and National Security Adviser Zbigniew Brzezinski. Turner told me that I would have exactly two and a half minutes of the meeting with the president. But Carter was confused about my identity and thought I was the "old hand," the man who had briefed Julio and myself the night before we flew into Tehran. That operative was still in Tehran, helping to prepare the military rescue operation. Once the confusion was cleared up, I showed the president some of the *Argo* cover materials, then more confusion ensued. The Oval Office photographer appeared, and Turner was uncertain if I was authorized to be photographed with the president, since I was working undercover.

Years passed before I would receive my picture shaking hands with Jimmy Carter. But I was thrilled to learn that I had been promoted to GS-15, the equivalent of a full colonel in the military, that very same afternoon by the Director of OTS.

◾

MY STOCK WITH the White House remained high that spring when Hamilton Jordan, Carter's chief of staff and most trusted confidant, asked me to prepare a disguise for him. Jordan had arranged ultrasecret European negotiations with Iranian Foreign Minister Sadeq Ghotbzadeh. If the meetings went well, the Carter White House would release

Iranian assets frozen in America in exchange for the hostages. But the slightest hint in the press that Jordan and Ghotbzadeh were secretly meeting would blow the deal. It was obvious that Jordan needed a fool-proof disguise.

I spent several hours with him in the White House basement barber shop, transforming the husky, clean-cut young man with the confident bearing of lofty authority into a rather frail, middle-aged gentleman who could walk unrecognized past his closest friends.

Later, Jordan would brag at singles bars around Washington that this disguise made him look like a "sleazy Latin businessman." In my profession, that was a compliment.

IN MAY 1980, Julio and I were awarded the Intelligence Star, the Agency's second-highest valorous decoration, at a secret ceremony in the Head-quarters auditorium. Other CIA officers were also being recognized for their involvement in the failed OPERATION EAGLE CLAW rescue mission, in which Sea Stallion helicopters that had secretly landed in Iran from an American aircraft carrier in the Arabia Sea had exploded, killing eight airmen. The classified nature of the ceremony was deemed necessary because the American diplomats still languished as hostages. They were not released until January 21, 1981, after 444 days in captivity.

Although I was honored to receive the award, my pride was bitter-sweet while Americans remained hostage in Tehran. I was equally dis-pleased that I was not allowed to invite Karen and the children or CIA friends to the ceremony, since the exfiltration of the six houseguests was still classified.

It was no secret to them that I had been in Iran during the rescue of the six. As with any Agency family, they were expected to endure my long absences with only scant details of the operations I conducted. Whether I was in Moscow or Tehran, we could not directly communicate, although the OTS fraternity did try to maintain an informal flow of information to our families to reassure them of our well-being.

For this reason, award ceremonies were important to Agency families. The presentations, usually held in the DCI's conference room with relatives present, acknowledged their sacrifice and the vital support they provided to the officer. However, Admiral Turner had decided the Agency needed its own morale boost following the EAGLE CLAW debacle and requested the closed ceremony in the Headquarters auditorium.

When OTS Director Dave Brandwein called me at home to announce the award presentation, I was about to leave on a twentieth-anniversary trip with Karen. "Tell the Admiral I'm out of touch," I told Brandwein.

He complied, but Turner issued an edict. "Find Mendez."

So it was that I received my Intelligence Star in the company of a few friends and relative strangers, while Karen and the children expressed their joy from a distance.

10 ▪ Endgame

> For anyone tired of life, the thrilling life of a spy
> should be the very finest recuperator.
>
> —Sir Robert Baden-Powell, British intelligence
> officer and founder of the Boy Scouts

Washington, November 1982 ▪ The street disguise exercise had gone well. My special OTS team had played the role of KGB bloodhounds, and the graduate career trainees from the Directorate of Operations had been eager to display the skills they had acquired in the IO (Internal Operations) Course. Even the *Mitteleuropa* weather—a nasty night of freezing rain that had cleared the streets of Georgetown and Foggy Bottom—had added to the realism of the experience.

The IO "pipeliners" had survived intense months of instruction, which had led seamlessly from the paramilitary discipline at the Farm to the more cerebral challenges of the advanced training, focusing on their areas of assignment. They were bright, high achievers, earmarked for important Soviet bloc jobs in their initial overseas tours. While "reading in" on the agent cases they would help manage, they were honing the tradecraft skills needed to operate effectively on hostile streets.

Although I had moved up to become Chief of Authentication in 1979, I remained personally involved in the disguise tradecraft training of these new officers and their spouses, and I took a special interest in those assigned to Moscow because I had helped refine the CLOAK–Silver Bullet procedures.

Transforming well-adjusted, law-abiding citizens into successful case officers was always a challenging, delicate process. By definition, the candidates had to be able to adapt to demanding overseas assignments. They had to be cunning and devious while working against the enemy, yet still retain their personal and professional integrity. Above all, they had to demonstrate an unwavering loyalty to their country and their colleagues, in that order. Beyond these seemingly inconsistent attributes, a case officer had to have that intangible flair, the ability to orchestrate complex exercises in deception tradecraft with ease.

When the pipeliners were cycled over to OTS for their postgraduate disguise training, I always stressed to our team that we had a serious obligation to fulfill in screening these new officers. Just because they had already passed muster for assignment to the Soviet Union didn't automatically qualify them to use the most sensitive Silver Bullet procedures on the streets of Moscow. These techniques were absolutely essential to our operations and highly vulnerable if misused.

Although we did not have formal veto power over a probationary officer's future, our colleagues in the DO's Soviet–East European (SE) Division, most of whom had been our comrades in the field, were certainly anxious for our opinions. Much of our evaluation of the new officers was visceral, based on an evolving doctrine and years of operational experience. Our instincts could tell us that one candidate would excel in the more specialized world of Soviet bloc espionage, while another, despite an impressive academic background, high language aptitude, and an engaging personality, simply would not. It was similar to the informal peer assessment soldiers had used for millennia, which boiled down to a simple question: "Would I trust my life (or an agent's) to this person?"

That night during the debriefing, one of the pipeliners, an attractive young woman who had been considered borderline because of her less proficient language skills, had used the CLOAK technique and completely eluded the surveillance team, disguising herself as a pitiful bag lady huddled on a steam grate near the Mall by adding a rain-soaked blanket and a tangled gray Halloween witch's wig she'd bought at a costume store.

"Way to go, Helen," I said, raising my can of Miller Lite.

My bloodhounds were taken aback but acknowledged that she had executed her escape flawlessly while still playing by strict "Moscow" rules of engagement.

Helen smiled radiantly, pleased by the first unqualified praise she'd received in weeks.

We moved on to the next officer, who had ignored the rules of engagement and was well aware of it. The man's name was "Darrell," one of the brightest of Soviet–East European Division's pipeliners. In fact, Darrell had been slated for a Moscow assignment early in his training process.

But I was disturbed by the report on his actions that night. One of the OTS surveillance specialists, "Jerry," who had been team leader on the exercise, had taken me aside just before the debrief. "The guy really became provocative between Washington Circle and Connecticut Avenue," Jerry had complained. "Then he doubled back toward the GW campus through a building in his DAGGER rig in order to make his next timing point."

Jerry's concerns were valid. That night's modified Moscow rules were a refined version of the specialized disguise materials employed in the CLOAK procedure, the new technique we'd code-named "DAGGER."

The DAGGER technique is still classified, so I cannot describe it in detail. But I can say that the disguise is so effective, it can be successfully employed while maintaining the flow of normal street travel.

Darrell knew this; we all did. Yet he had chosen to break the rules. Why? He already had his coveted Moscow assignment. Was his competitive drive too strong to control?

I studied Jerry's handwritten surveillance notes, then turned to Darrell, a dark-eyed, calm, and self-assured man who looked younger than thirty-one.

Maybe tonight's breach was an aberration. Or perhaps he hadn't fully understood the rules of the exercise. I handed the report to the Soviet–East European control officer, "Martin."

"Darrell," Martin said, looking him directly in the eye, searching for any sign of deceit, "Jerry saw you duck into a building when you doubled back up L Street and then into that alley near the *Washingtonian* office. That was against the exercise parameters. Why did you do it?"

Darrell hardly blinked. "I didn't. Once I lost surveillance, I just kept moving toward Connecticut until I hit my timing point and changed into the DAGGER rig."

"We *saw* you," Jerry said harshly.

"If you ran out of time," Martin said, trying to control the palpable animosity building around the table, "just admit it, and we'll repeat the exercise another night."

Darrell coolly looked each of us in the face before speaking, then returned his gaze to Jerry. "You're wrong."

He's trying to "case officer" us, I thought. Lying was bad enough. But trying to outwit us through this transparent deception with his colleagues was a worse offense. Suddenly, I remembered that painful morning in 1967 when Lynn, my Flaps and Seals instructor, had caught me in my lie about using the forbidden French opening to unseal her test envelopes. Then, I had been humiliated and condemned to clean her lab for a week, but I had also learned a valuable lesson about mutual honesty among colleagues in this strange business.

"Okay," Martin said flatly to Darrell, then turned to the next trainee. Perhaps the embarrassment we all felt for Darrell would shame him into admitting his deceit. But he sat serenely at the table and even managed to eat a slice of pizza as if nothing had happened.

Jerry looked at me quizzically, but read my expression correctly. I would consult with Soviet–East European management about this situation later, an unpleasant but necessary obligation.

But I didn't have to make a special trip to give my report. Mary Peters, the Moscow case officer who had been arrested by the Seventh Chief Directorate and PNG'd from the Soviet Union following TRIN-ITY's roll-up, had been overseeing Darrell and other pipeliners in their advanced tradecraft. She came to see me to voice her own concerns.

"We're getting some unusual reports on Darrell," Mary confided. "Nothing earth-shaking, but still disturbing."

"The only definite thing I have is a lie he told last week in the after-action session," I explained. "But the guy just stuck by the lie, even when Martin nailed him with it. I don't like that, Mary."

"Neither do I," she agreed.

It didn't take long for cocky Darrell to step down from his high horse. As all the older, more experienced case officers recognized, the pressure and stress of this intense training, as well as the unnerving prospect of a pending assignment to the big league of a Soviet bloc station, had caused the young pipeliner to falter in his basic integrity. When he was again confronted with his performance in rigorous tradecraft exercises, Darrell came clean and asked to be given another chance.

He had learned a painful but precious lesson: Espionage was an extremely stressful business, a profession that combined elements of being a street cop with those of a salesman working on commission. In Darrell's case, his progress through the pipeline was somewhat slowed, and he was given a couple of extra months to prepare himself for the field. This grace period allowed him to mature enough to handle the stress under which he had almost buckled that rainy November night.

Blue Ridge Mountains, May 28, 1986 ■ I stood on the rutted track between the house and the cabin in a natural alcove formed by black locust trees whose branches were overgrown with wild climbing roses and honeysuckle vines. The fragrance was almost overwhelming. I had come here this bright spring Saturday morning just after sunrise to pick a bouquet of wild flowers from our forty acres to be placed on the altar of St. Luke's Episcopal Church at Karen's memorial service.

Suddenly, I sensed that I was not alone, that there was some intangible presence surrounding me in the trees and vines. As I stood there motionless and listened, all my senses sharpened. Then I walked slowly back toward the house, taking measured steps—I didn't want to look as if I was fleeing that unknown observer.

Almost as quickly as the feeling had come it disappeared.

Grasping the bouquet, I reflected on the series of near-mystical events that had occurred during the week since Karen's death. I had visited that wooded alcove several times and had always been aware of that inexplicable presence. The night before, I had invited my son Toby, and his sister Amanda, to join me there. Toby had returned from school in Chicago two days earlier and Amanda had just arrived from Seattle. Neither had been with Karen in her final days.

Karen had been diagnosed with lung cancer in January and received radiation treatments at Johns Hopkins Hospital until April, when she appeared to be in remission. Her sudden relapse and death had come so unexpectedly that there had not been time for the two oldest kids to return home. But I had brought Ian, our youngest, from Washington to be with Karen the day she died. The trauma of her disease and death had left all of us reeling with grief.

On Karen's fortieth birthday, I had surprised her with a hot-air balloon flight over our valley, something she had wanted to experience for years. This small extravagance hardly compensated for countless missed birthdays and anniversaries, when professional assignments had pulled me away from her and the children. When I had been in the mountains of Laos, on extended duty in South Asia, or in the Soviet bloc, I had often thought of our life together in this Blue Ridge retreat after retirement, when I would finally have the chance to paint full time, which had been one of our naive dreams as a young couple in Denver.

Two years after her fortieth birthday, when she could hardly bear the pain of her cancer, Karen's memory of that balloon flight lifted her from the agony she faced, and now, on the afternoon of her memorial celebration, we had asked the same balloonist to take up her ashes and scatter them over the wooded hillsides she had loved so deeply. The balloon was launched from behind the house at sunset with a hissing roar and the brilliant glare of its propane burners. A crowd of family and friends watched from the deepening shadows on the ground as the multicolored balloon rose into the sky. When the balloonist finally threw open the urn, the ashes showered across the land in sparkling pink clouds of sun-kissed rain.

CIA Headquarters, Langley, May 1986 ■ Two weeks before Karen's death, a case officer named Aldridge "Rick" Ames, serving as the DO's Chief of Soviet and East European Counterintelligence, faced the prospect of a polygraph examination before leaving on assignment to the Rome station. Although the test was routine, Rick Ames was worried about being "boxed."

Ames, the son of an Agency officer, had served with the CIA since 1962. Rick Ames's assignments had included Ankara, Turkey; Mexico City; New York (where he attempted without success to recruit his Soviet counterparts at the United Nations); and several years at Headquarters.

In all his jobs, Ames had turned in a mediocre performance. He had also become an alcoholic; suffered through a costly, acrimonious divorce; and fallen hopelessly into debt. But his second wife, Rosario, a junior Colombian diplomat he had courted in Mexico, had extravagant tastes and was devoid of any loyalty to the United States.

A tall man with sloping shoulders and thick glasses, and an acerbic arrogance, Ames soon succumbed to Rosario's desire for the "good life" they had known living under embassy cover in Mexico.

In April 1985, Ames had simply walked into the Soviet embassy two blocks from the White House on 16th Street NW and volunteered to sell his services to the KGB. By way of bona fides, he presented his Agency ID card and Headquarters pass, as well as copies of several sensitive Soviet–East European Division cables.

In the months following Ames's brazen walk-in defection, he sold the KGB the names of every important CIA asset in the Soviet bloc. During those months, the KGB relentlessly pursued our agents-in-place, rolling up over forty assets.

One of the cruelest losses was the arrest and execution of GRU Lieutenant General Dimitri Fedorovich Polyakov, who had served as a vital agent-in-place since being recruited in New York in 1962. Polyakov was the most senior Soviet military intelligence officer to provide information to the CIA. Following his arrest in Moscow and brutal interrogation, he was executed in 1986.

By the end of his first year of espionage, Ames had cleared his debts, and he and Rosario embarked on an outrageous spending spree. He was soon making recklessly large deposits, often transporting paper bags of cash directly from Washington-area dead drops, where the KGB had stashed them, to the drive-through window of his bank.

Fortunately for Rick Ames, as the catastrophe in the Soviet bloc unfolded, the natural supposition at both the Moscow station and Headquarters was that this string of betrayals could be attributed to a known traitor, Edward Howard. A disgruntled former pipeliner who failed a polygraph about drugs and alcohol abuse on the eve of his departure for Moscow, Howard was fired by the CIA in May 1983. Within one year he had volunteered his services to the KGB and sold the Soviets the names of several agents he had been trained to work with in Moscow. Finally, tracked by the FBI, Howard had defected to the Soviet bloc in

1985. Equally devastating, he sold the Soviets the details of our Silver Bullet disguise techniques.

Counterintelligence assumed that Howard, as a gung-ho career trainee, had managed to become more knowledgeable about the Soviet bloc cases than anyone had ever imagined.

That false conjecture had protected Rick Ames throughout his first hedonistic months of treason. Now he had to face the polygraph.

His KGB case officer, Washington *resident* Stanislav Androsov, tried to ease Rick Ames's mounting panic. "Try to get a good night's sleep, and relax," he urged.

Ames followed this advice and used disassociation techniques to evade standard questions about contact with foreign intelligence services and the divulging of classified information. The fact that he felt no remorse over the Soviet bloc agents' deaths undoubtedly helped him to survive this examination, as well as another routine polygraph in 1991.

Although Ames's May 1986 polygraph did reveal possible signs of deception, his long tenure in sensitive CIA positions served to protect his squandered integrity. Unlike Edward Howard, Rick Ames was old-line Agency family. No one in Counterintelligence would believe that he had sold out either his country or the CIA. In addition, because Rosario came from a prestigious Colombian family, Rick used his case officer skills in deception to his advantage, convincing colleagues that he had converted a nest egg from her parents into a small fortune, benefiting from stock market tips given by an unnamed old college buddy.

With luck on his side, Rick Ames continued his treachery unabated throughout his Rome tour and on his return to Headquarters. He had become the nightmarish "mole" that the Agency's archetypal cold warrior Chief of Counterintelligence, James Jesus Angleton, had hunted in vain for almost two decades.

Eventually, the KGB paid Ames almost $3 million, making him by far the most expensive agent in history on the Soviet payroll. Ames and Rosario paid $540,000 in cash for a lavish home in an exclusive northern Virginia suburb, and he brazenly parked his $41,000 fire-engine-red Jaguar XJ6 in the Langley parking lot.

By the time the FBI finally arrested Ames in 1994, the Cold War had ended. He pleaded guilty to espionage under the National Security Act and offered to cooperate fully, provided that Rosario received a lighter sentence. A federal judge sentenced Ames to life in prison without parole; Rosario received less than six years. At least $100,000 of the KGB payroll had disappeared into Latin American banks and to this day has not been retrieved.

Besides General Polyakov, nine other major Soviet bloc agents-in-place were executed and dozens were sent to the Gulag. Equally devastating, from the perspective of American intelligence collection, was that the KGB "doubled-back" an unknown number of the other compromised agents, who became sources of tainted information. Ironically, the disaster of the Rick Ames betrayal led the CIA to shift its operational emphasis in the Soviet bloc from clandestine agent intelligence collection to covert actions in order to bolster the mounting political resistance movements in Eastern Europe. During this endgame of the Cold War, I would be called upon to play an important role, which drew on the same artistic background that had originally led to my Agency recruitment.

In July 1986, I was promoted to the Agency's Senior Intelligence Service (SIS-1), the equivalent of a military general, and one of the federal government's elite Executive Rank employees.

Foggy Bottom, October 1986 ■ "Go see Casey." Peter Marino, the Director of OTS, stood in the door of my Central Building office, having just come from a meeting with his boss, Evan Hineman, Deputy Director for Science and Technology.

"When?" I asked, wondering what urgent task or sudden operation would require my meeting with Bill Casey, Ronald Reagan's mercurial and controversial Director of Central Intelligence.

"Now," Peter said. "Today. Casey wants to see Agency officers with promising ideas."

Jesus, I thought, here was the chance of a lifetime for an old Pinball player like me to rack up some points, but Casey wasn't giving me time to put on a carefully rehearsed dog-and-pony show.

"I'll take the next Bluebird," I told Peter.

"Take your car. He wants you there ASAP."

Fighting through the midday traffic on the northbound George Washington Parkway, I considered what might lie behind Casey's urgent request for a meeting. The recent debacle of agent betrayals in the Soviet bloc had been a successful KGB counter to our string of Silver Bullet victories, but the Agency had not yet unraveled the secret behind those betrayals. Speculation ran the gamut from Edward Howard's treason, to the possible compromise of our global communications network, to the physical penetration of the Moscow embassy's inner sanctums (perhaps with the compliance of two corrupt Marine guards, corporals Clayton Lonetree and Arnold Bracy, who had been ensnared in a KGB sexual "honey pot"), to the existence of a high-level mole in the CIA Headquarters, and, finally, to a nearly complete breakdown of our carefully established Moscow tradecraft. A Special Task Force had been established to investigate each of these possibilities.

The Task Force's inquiries slowly proceeded with the harsh knowledge that the Soviet–East European Division's clandestine intelligence, collected through recruited agents, would be suspect until the Agency found answers, meaning that our efforts in the Soviet bloc would have to shift toward other operational means. Many of us had definite ideas about what those means should be and the technical methods we should employ to achieve our ends.

As the new Chief of the Graphics and Authentication Division, one of my most pressing objectives was to modernize the Operational Graphics side of OTS, which I felt was a substantially underused resource. Graphics, like the disguise capability twelve years earlier, had seen better days. Not only had much of the Agency's old talent retired, outside, overt graphics technology had progressed remarkably in that period and passed the Agency by.

These advances presented us with both amazing opportunities and daunting challenges. The use of computers to digitally reproduce both text and image meant that my Division could more effectively support alias document and covert action propaganda operations throughout the Soviet bloc. These operations ranged from agent rescues to supporting the banned Solidarity trade union's underground newspapers and similar anti-Communist material that the Czechoslovak resistance group Charter 77 circulated. Such operations were a natural extension of the decades-long GIDEON program, in which OTS/Graphics had produced miniature Bibles, which were then smuggled into Eastern Europe by the DO. This clandestine campaign had also included publishing miniaturized copies of works by Soviet and East European dissident writers such as Boris Pasternak, Alexander Solzhenitsyn, Andrei Sakharov, and Czech playwright Vaclav Havel.

When I was a young trainee in the Graphics bullpens working on covert action propaganda, producing some of these materials would have taken months using traditional design and printing techniques, and would have presented major logistical problems for their secret placement in Eastern Europe. But by the mid-1980s, a three-and-a-half-inch floppy disk carried in a "mule's" hidden pocket could hold all the digital data an underground operator needed for a job. Using a basic computerized printer and a copy machine, the operator could produce anti-Communist flyers in Hungary or posters announcing Solidarity rallies in Warsaw.

The digital revolution clearly offered the CIA the opportunity to transform its propaganda effort in the Soviet bloc from a relatively marginal operation to a major campaign—*if* Casey and the Seventh Floor continued to back us. But the challenges inherent in computerized graphics could not be ignored. Around 1980, foreign national identity cards and travel documents began to incorporate new security technology in both the printed material and the computerized scanners at airports and border crossings.

The growing use of such digital security controls in the West would soon spread to the Soviet bloc, led by the Czechs and the East Germans, and would threaten CIA alias document operations across international borders. My predecessor had recognized both the opportunity and challenge that information-age computerized graphics represented and had launched an initiative to secure expensive, high-powered computers and the innovative digital scanning equipment needed to produce major document reproductions.

But this revolution had met with stubborn resistance among the old hands of Central Building. Invention of the digitalized reproduction techniques was in the care of one bright individual, whom the old guys in the art bullpen, the photo/plate lab, and the pressroom considered a threat to their craft and their jobs.

After I took over the Division, I saw this dispute as a cultural resistance to change, a bottleneck in the inevitable route to technical progress. I bombarded my superiors with program plans describing a broad-based effort that would draw on the power of new technology to create a separate digital graphics section, dedicated to both the opportunities and the challenges that the information-age revolution had given us. In putting forward my plans, I drew on the eighteen months I'd just completed as Chief of the Clandestine Imaging Division, in which we'd revitalized the electro-imaging program. Our boldest initiative had been a planned technical penetration operation directed against the Kremlin and several Soviet ministries in Moscow, using innovative new technology to gain entry into Soviet information systems.

We held a series of short briefings on that program for key CIA leaders in the EXCOM (a handful of Agency officials of the highest level, chaired by CIA's Executive Director). The outcome of these briefings was many millions of dollars in D&E (Development and Engineering) funds, which would be used to build the systems needed to complete the technical penetration operation.

Pulling into the main gate of CIA Headquarters off Route 123, which I had first entered as an apprentice artist in 1965, I hoped that the notoriously irascible William Casey would find enough merit in our achievements and plans to continue funding the computer revolution under way in my Division.

"PLEASE SIT DOWN, Tony," Bill Casey said. He spoke with dry precision and refrained from the incomprehensible mumbling that so often marked his congressional testimony.

Casey, a bald, shuffling bear of a man in an expensive pin-striped suit, motioned my two branch chiefs, "Daniel Morgan" and "Dennis Norman," whom I'd invited to the meeting, to sit on a plaid sofa in the corner of the suite. When I turned to join them on the sofa, Casey gripped me by the elbow and eased me into the wing chair beside his desk. He moved forward in his swivel chair so that our knees almost touched.

"I'm here to listen," he said, his intelligent, lively blue eyes probing me from behind wide designer glasses.

Although I'd faced more than my share of hostile surveillance, Casey's benevolent scrutiny was surprisingly unnerving. My apprehension was no doubt due to his legendary reputation. As General "Wild Bill" Donovan's closest OSS subordinate in Europe during World War II, Casey had wielded incredible power, mounting covert operations and dispatching Allied officers and recruited German agents to carry them out. In the four decades since, he had become a millionaire Wall Street lawyer, served as president of the Export-Import Bank, and kept his hand in espionage as a member of the President's Foreign Intelligence Advisory Board during the Carter and Ford administrations. After successfully running Ronald Reagan's 1980 presidential campaign, Casey had been a natural choice for DCI.

What most impressed Agency officers early in Casey's tenure at Langley was his virtually unlimited access to the president. In my two-week Senior Intelligence Service "executive charm school," which I attended after being promoted, I was told informally that Casey was "the best DCI since Allen Dulles. He's elevated the position to cabinet rank."

However, patience was not Bill Casey's strong suit. He believed in results, and he believed in delivering success to his president.

But in the eleven years since the fall of Saigon, success had been conspicuously absent for the West in the Cold War struggle. Soviet surrogates had flourished in Southeast Asia, the Middle East, the Horn of Africa, and Portugal's decayed African empire. At the time of my meeting with Casey, the Soviets had established a firm beachhead in Nicaragua, and the Marxist guerrilla army in El Salvador was resolved to overthrow the corrupt military-dominated government of the oligarchs, whom we had been forced to embrace as our allies.

Congress, through the Boland Amendment, had effectively prevented the CIA from directly arming and training the *Contra* peasant rebels in their battle against the Soviets' Nicaraguan surrogates, the

Sandinista government. But as the world would soon discover, Casey had bulled his way ahead, helping evolve the Iran-Contra arms supply scheme, one of the most convoluted, and ultimately wrong-headed, covert operations of the Cold War. In an elaborate deception operation, the CIA sold TOW antitank missiles and other munitions to Iran, embroiled in an endless war with Iraq, which then prevailed on its Shiite allies in Lebanon to free American hostages on a piecemeal basis. Casey's National Security Council protégé, Marine Lieutenant Colonel Oliver North, then used the proceeds from the missile sales to buy arms for the Contras.

Should the CIA have been involved in such a wild scheme without direct congressional approval and the formal authorization of a presidential "finding"? Of course not. Could anyone at Langley have stopped Bill Casey once he had made up his mind? I doubt it.

Now I had to convince this unpredictable genius to continue backing the efforts to modernize our clandestine graphics operations.

As succinctly as I could, I explained my "proactive" plan to prepare the Agency for the worldwide proliferation of computerized border controls and the threat of more sophisticated personal identity and travel documents, which were beginning to appear in both the East and the West.

"I think we can make inroads against these threats by helping to lead the industry in the right direction," I said. "State and INS want to include us in open symposia. This research activity would be in the public domain. We have friends in academia who can help lead our efforts."

Casey's nimble mind immediately grasped the implication of my proposal. Once the United States helped lead the world's experts on computerized security controls and high-tech documents, Soviet bloc spies and terrorists would find operating across borders more challenging. If the Agency acted swiftly, we wouldn't be caught out in the cold when a new generation of technology quickly emerged, as it always did, and we wouldn't have to reverse-engineer in order to catch up.

"What do you need from me?" Casey asked calmly, giving his implicit backing to the project.

I knew if I asked the DCI for ten million dollars, he'd probably offer fifteen. But I wasn't greedy.

"Nothing for the moment, sir," I said. Then I pointed to Dan Morgan, a bright photo scientist I had made Chief of Graphics Production, and Dennis Norman, who I had just named Chief of Operations for my Division. "But as soon as we have something concrete, we'd like to come back and see you."

"Any time," Casey answered.

But time was something Casey did not have. In two months, he was embroiled in the unraveling Iran-Contra scandal. Stricken with a brain tumor in January 1987, he died within weeks.

However, we eventually received adequate funding for our ambitious plans.

"WELCOME TO THE Magic Kingdom, Mr. Vice President." "Roger," the chief of our computerized graphics "crew," wearing Mouseketeer ears, pinned a commemorative badge on Vice President George Bush when he entered the new Central Building operation in April 1987.

A former Director of the CIA, Bush was still fascinated with intelligence and was engrossed in our briefing on Graphics Modernization.

His eye was caught by a modem scanner imaging device that allowed us to inspect suspected terrorist passports remotely over telephone circuits from around the world. Cooperating security services could simply scan the passport at an airport station, and the image would be instantly received for analysis at the CIA's new counterterrorism center at Langley. This new technology would be deployed at the forthcoming Olympic Games in Seoul.

"What are the other applications of this system?" Bush asked.

Roger explained that we were working on a secure digital imaging network of high-resolution images and exemplars of documents for our analysts at Headquarters and our case officers in the field. In the past, such materials would have been hand carried by courier.

Roger held up a "music" cassette, indistinguishable from millions of others found worldwide. But when he inserted it into a computerized playback deck of the type we had installed at several locations behind the Iron Curtain, images of foreign text scrolled across the monitor.

"We also send material on floppy disks and the hard drives of PCs," Roger explained.

"The more personal computers proliferate in the world, sir," I told Bush, "the more porous borders have become. Within a couple of years, my guys tell me, we're going to have computers that fit on your lap but can hold the entire text of the *Encyclopaedia Britannica*."

"Really?" Bush said politely, but he did not seem convinced.

Moscow, July 1988 ∎ The predawn glow was faint through the soot-covered window of the apartment compound garret. I slipped out of bed and dressed quickly, finding my clothes in the gloom where I had put them the night before. Less than two minutes later, I was stealing down the three flights of stairs, avoiding the elevator, which would alert the KGB that someone might be about to leave the building.

To my amazement, there was no militia man in the guard shack at the archway entrance on this Sunday morning. Things had certainly changed in the twelve years since 1976.

I moved down the tree-lined sidewalk, turned onto Kalinina Prospekt, and across the Kalininsky Bridge. At this early hour, the streets were absolutely empty. *Where's surveillance?* I wondered. I had chosen this time and route, a provocative variation from my usual morning jaunt, hoping to stimulate some KGB activity. But if anyone was following me on foot or vehicle, they truly had mastered the art of being invisible, as some officers in the Moscow station were beginning to believe.

On the other side of the bridge, I lengthened my stride, skirting the grotesque spires of the Ukrania Hotel and entering a side street that led toward the Kiev railway station and the Kievskaya Metro stop. If there was distant surveillance on me this morning, the team would not want me to enter the train station or Metro unaccompanied. Approaching a wider street, I descended the stairs to a pedestrian underpass. Now I was sure I had complete privacy because no one could see me for several minutes.

Off the sidewalk beyond the underpass, a gate opened to the right through an evergreen hedge. Passing by the night before, I confirmed that the hedge surrounded a small, secluded playground reserved for the children of the *nomenklatura,* living in the high-rise apartments that dominated this exclusive neighborhood near the embankment. This postage-stamp playground was not at all like the muddy, rubble-strewn lots near the *Khruschoba* prefab apartments, thrown together in the 1970s in the industrial districts of Moscow. Here, the grandchildren of the Party elite enjoyed brightly painted swings, slides, and a well-varnished teeter-totter.

I turned into the playground, closed the gate behind me, and waited.

Standing close to the hedge, I could not be seen from any high-rise window with a view of the playground. I waited five minutes, but no one walked by the gate. Then I waited five minutes longer. Still, I had no company. I sat down on a cast-iron bench shielded by the hedge and reviewed the events of the past ten days in Moscow.

Jacques had requested I join him here, where he had been reassigned by the Deputy Director of Operations, Clair George, to see if he could restore CIA operations, which had been practically shut down after the Howard defection and the still puzzling 1985 mass roll-up of our agent network. The cause of the debacle still remained a controversial mystery, with some factions willing to assign blame to ill-defined "sloppy tradecraft" on the part of the Station or our agents-in-place.

Most of the people I worked with, however, understood that the failure of street operations in the Soviet Union could not have triggered all of the agent roll-ups of the mid-1980s. Several of our assets had not even been inside the Soviet bloc when they were compromised but rather had been operating at Soviet embassies in the West and had suddenly received orders to return to Moscow for "consultations." Wisely, they had defected.

Nevertheless, the Agency wanted someone like Jacques to sift through the confusion, to find a new way of doing business, and to invent new Moscow rules. My involvement in developing the original Silver Bullet led Jacques to ask me to help his station invent another.

Over the past ten days, I had sat listening to each case officer relate how the KGB had counterattacked with a vengeance, exacting a heavy price for the damage we had inflicted on them during the eight years the Silver Bullet went unchallenged.

"They're intent on closing down every damn one of our operations," one officer complained. "But they also want to humiliate the CIA whenever they can."

Many of the officers newly assigned to Moscow didn't know whether such retaliation was simply meant to unnerve them. Others thought the KGB was luring us into complacency through phony breaks in surveillance, in which the Seventh Chief Directorate teams stayed back over the visual horizon for hours on end but could still converge the moment an officer decided he was clean and went operational.

"They've got some kind of new 'ghost' surveillance," one officer whispered to me, even though we were in the secure confines of the Bubble. "They're flaunting it too, Tony, which has everybody demoralized. They couldn't keep shutting us down like this unless they had some kind of unseen advantage."

I had never felt such an unsettling atmosphere in Moscow. It only got worse with the sudden redefection of KGB Colonel Vitaly Yurchenko in November 1985, when he had slipped away from his inexperienced CIA security guard at a Georgetown restaurant and walked up Wisconsin Avenue to the new Soviet embassy compound. Many in the Agency were convinced that Yurchenko had never been a true defector, but simply a brazen plant meant to dangle Howard's name as a deceptive lure in order to mask the true source of the disastrous betrayals, maybe even to cover the tracks of a KGB mole operating at a much higher level in the CIA.

So I was back on Moscow's streets as an operative trying to make sense of this new environment. But, even though I had only ten days to consider the problem, certain facts did not add up. If anything, the overall operational situation had actually loosened in Moscow. Glasnost was obviously real, not mere propaganda. I'd seen people sitting in Gorky

Park reading the new independent investigative journals such as *Argumenti i Facti*, which were dedicated to exposing both official corruption and the Soviet Union's dark past, ultra-taboo topics when I had last been in Moscow. A newly organized private organization, "Memorial," documented the fates of millions of innocent Soviet citizens, who had been swallowed by the abyss of the Gulag. The Kremlin had even stopped jamming the Agency-funded Radio Liberty's broadcasts.

Although perestroika had not brought the miraculous economic changes Gorbachev had hoped for, Moscow itself had lost much of its gray veneer and acquired a more prosperous appearance. I had seen a fair number of Volkswagens and even some Volvos on the streets. People were dressed in more vibrant colors, and they seemed hopeful. There were now privately owned restaurants where one didn't have to have the *blat* of Party connections or hard currency to get a table.

After one Sunday afternoon promenade down the Arbat pedestrian mall, where I'd encountered thousands of Muscovites dressed as fashionably as their counterparts in Frankfurt or Stockholm moving among the vendors hawking jewelry, leather goods, and tourist kitsch, I had stood on the corner near the Foreign Ministry and counted cars with diplomatic or foreign commercial plates at five-minute intervals. It was impossible for the KGB's mobile surveillance to cover all these vehicles as it had in the past. At the Intourist Hotel, I counted twenty-three Mercedes, Peugeots, and Opels with West European plates, proof that Westerners were taking motor vacations in the Soviet Union. The campgrounds dotting the wooded suburbs were full of German, Dutch, and Belgian vans.

Perestroika, and the desperate need for hard currency these tourists carried, had almost completely undercut the KGB's ability to effectively seal off the streets of Soviet cities. But surveillance still existed. The day before, Jacques asked me to accompany a CIA communicator in his car and on foot during an extended shopping outing. It took us a couple of hours, but I finally identified a discreet mobile surveillance team of six or eight men and women in a variety of clothing styles who hung back while maintaining full coverage. Then, the night before, I stood across from the Lenin Library on the crowded sidewalk of Kalinina Prospekt and observed the intricate ballet of a crack surveillance team scrambling to catch up to a bolting rabbit, who had apparently disappeared down one of the numerous Metro escalators.

On my first trip to Moscow, I had wondered if the widespread resentment over the Soviet Union's corrupt and stagnant economy would one day undermine the KGB's greatest strength—a submissive, even fearful, population. Now I realized that glasnost and perestroika had eroded the resignation and apprehension of the Soviet people. They wanted a better life. They wanted honest leaders. They wanted the freedom and opportunities we had in the West.

Had I been a betting man, I would never have laid money on the KGB's ability to keep holding down the lid on Soviet society, as it had for so many decades. Times had changed.

Where is their surveillance now? I wondered, glancing around the empty playground. Had my alias legend held up, despite the previous day's outing with the CIA communicator? Could we still create elaborate deceptions based on an expanded "persons of little interest" group?

These were hotly debated issues within the Station. Some officers believed that even a very discreet stand-off surveillance was merely the Soviets' "B Team," while there was always an invisible "A Team" out there who would only appear at the most compromising moment for officer and agent. But was this physically possible?

Despite the spreading thaw in the Soviet bloc, that question remained vital to our future operations. The Cold War had taught us that détente could quickly give way to a deep freeze and military confrontation. I deeply wanted to believe that the end of the struggle was in sight, but I had to fulfill my professional responsibilities.

In my list of conclusions and recommendations for Jacques and his officers, I would describe my plan to create my own A Team in Washington, a spin-off of the OTS street-disguise training exercises. Using volunteers, I would test the most up-to-date varieties of our "portable" and adaptable disguise technology, which held the key to our future street operations. Once we were sure that we could operate against any type of ghost surveillance the KGB might have perfected, we could begin writing a new set of Moscow rules.

It was time to go. I decided to leave the playground through the only exit, the gate in the hedge. My operational senses, honed over twenty years in the field, told me that I had been alone on that silent bench. But if I had been watched, the KGB would have searched every inch of the playground for drops and staked it out for weeks or months to come.

Strolling back over the river on the Borodinsky Bridge, I thought about the challenges facing both intelligence services. As the Soviet bloc opened its doors to the West, the KGB would be stretched beyond its logistical and financial limits. But its grudging respect for the CIA would compel it to remain in no-quarter combat with us until the end of the struggle.

CIA Headquarters, October 1988 ■ Burton Gerber, Chief of the Soviet–East European Division, sat behind his desk in the corner suite on the fifth floor of the Original Headquarters Building, apparently listening impatiently. I was trying to make a coherent presentation of my conclusions and recommendations from my Moscow survey. I had already submitted my report by cable, but I was interested in both his insights and his attitude toward my approach. In my opinion, the problem was that Burton seemed too self-absorbed to acknowledge any operational approach other than his own. The man I remembered, preparing for a Moscow assignment several years before, had become a complete stranger now that he was back as Chief of the Division. Once very receptive to our suggestions and training, his senior assignment had transformed him.

"Things are changing in Moscow. Something new is definitely in the wind," I told Burton. "Everything from the clothes people are wearing to the optimism you can actually see in their eyes. Moscow's not the same city it was in 1976."

"Ridiculous," Burton said scornfully. "I was there this year myself and saw nothing of the sort."

The tone of the entire meeting was unpleasant and contentious. Every argument I presented, Burton contradicted. I sensed that he and Gus Hathaway, Chief of Counterintelligence, were under tremendous strain following all the Soviet bloc betrayals. I also suspected that they were in general conflict with Jacques, since both Burton and Hathaway had recently been chiefs of station in Moscow and were probably second-guessing everything Jacques did.

One of the fundamental assessments emerging from Jacques's Moscow station in 1988 resonated with my own evaluation: The drab predictability of post-Stalinist totalitarian existence was crumbling. Boys of eighteen from collective farms no longer marched proudly to the army recruiting stations, accompanied by their patriotic parents. Instead, draft evasion was widespread. Once in the army, young men did everything possible to avoid combat service in Afghanistan. The sons of the *nomenklatura* simply paid off doctors for medical exemptions.

This kind of strain was predictable enough as the Soviet empire rumbled toward an uncertain future. But glasnost was a wild card. A little freedom went a long way. State Television had recently unveiled *Vzglyad* (Glance), a *60 Minutes*-type news magazine that seemed hell-bent on exposing every scandal hidden behind the Kremlin walls.

Burton, however, was not impressed with these developments. I let my mind wander as he held forth about Moscow in direct contradiction to my views. I felt a sudden chill listening to Burton's rambling. It seemed to me that he had lost the most important attributes of an operational intelligence officer: practical flexibility and a neutral perspective. To succeed, we had to accept the world as it was, not as it had been or the way we wished it to be, even if an upheaval of enormous proportions had occurred, as seemed to be the case in the Soviet Union.

Yet Burton could not accept this strange reality. In my opinion, he was a man in denial.

The White House, Spring 1989 ■ Jonna Goeser, who had recently moved up from Deputy to Chief of Disguise, had been one of my strongest collaborators in developing the OTS Special Surveillance Team's variant of the KGB's "ghost" surveillance in Moscow. But we had gone one step further, incorporating into these procedures a variety of extremely refined DAGGER disguise techniques, which remain classified.

As one of our best officers, it was only fitting that she be given the honor of unveiling this disguise technique to President George Bush. The Oval Office in the White House, of course, was not a place that many Agency officers visited, aside from the DCI and very senior executives.

I knew of only three OTS officers who had been there: myself after the Tehran operation in 1980; Dr. Crown, Chief of the Questioned Documents Laboratory, who briefed President Carter on Soviet disinformation; and Jonna, who accompanied our new DCI, William Webster.

Earlier that morning, Jonna had driven to Judge Webster's suburban home to prepare for the innocent deception operation they planned to unleash in the Oval Office. On entering Webster's front hall, she was practically attacked by a snarling little terrier. But when Jonna emerged from the ground-floor powder room in her disguise, the very latest improvement on the earlier DAGGER, the little dog wagged its tail and sat up sweetly to be petted.

When she and Webster arrived at the Oval Office anteroom, several senior National Security aides, including Deputy National Security Adviser Robert Gates, were waiting. The wait stretched to almost forty-five minutes, with no one guessing what was about to occur. All they saw was Judge Webster standing beside an attractive, young, and animated aide whom they did not recognize.

Once the group was seated before President Bush's desk, Webster introduced Jonna with a pseudonym, describing her as an "Agency specialist" who had very urgent information for them.

Jonna rose and opened her briefing book. "Mr. President . . ." she began.

George Bush seemed intrigued, then disconcerted, as Jonna reached up to her face and pulled away something . . . Suddenly, the diminutive young brunette seemed to slip away. The Secret Service guard at the entrance door moved forward, but Judge Webster raised his palm to stop him.

Jonna had metamorphosed into a classy lady of forty, with untainted makeup and tasteful jewelry. As she laid down her briefing book and tossed the paper-thin disguise into the air, it hung suspended in the sunlight for a moment before she caught it.

The leaders in the room were speechless.

New Headquarters Building, Langley, Summer 1989 ■ "The Estonians are planning to print their own currency," a well-placed corporate source called to tell me. The news was stunning, especially as "Steve," my immediate OTS supervisor, soon personally verified its reliability.

Our operation had moved into the New Headquarters Building at Langley, and the explosive intelligence revelations thundering out of the Soviet bloc reached us much more quickly than when we were based in Foggy Bottom.

One of my first acts after hanging up the telephone was to visit the young officer I planned to send into Estonia on a clandestine probe operation. The Baltic republics were tinder dry, ready to flare into open defiance of Moscow as had Soviet Georgia earlier that spring, only to suffer a sickening massacre of unarmed civilian demonstrators in the capital, Tbilisi. So I had to caution this young officer.

But when I offered my information about the Estonian intentions to the Soviet–East European Division, I found that they had gone from denial to disbelief.

"The KGB will never let them dump the ruble," the Chief of SE/ Internal Operations told me. "We have no contingency plans on how to handle this kind of situation."

Events in the Soviet bloc were moving too fast for many in the Agency to exploit. But I realized that conditions in the Baltic states rep- resented a unique opportunity for us to lay down a network of exfiltration routes. That summer, we dispatched two bright young members of our Special Surveillance Team to do a recon.

Reporting back in September, they described three virtually independent little nations that had turned their backs on Moscow. The medieval Baltic capitals, ringed by shoddy ranks of *Khruschoba* prefab apartments, had become virtual enclaves of the West.

"They'll never go back to the old system," the young officer told me.

"No way," her partner echoed. "The genie's out of the bottle, big time."

It might not have been the type of field assessment the old-school analysts in the DI liked to hear, but, despite the youthful exuberance of the two officers, their message rang true.

CIA Headquarters, Langley, November 9, 1989 ■ The television in my outer office was tuned to CNN's live coverage of the incredible events in Berlin. As the camera zoomed in for a closeup, a husky young man in a rock band T-shirt swung his sledgehammer with one final blow, and a wide, graffiti-plastered section of the Wall toppled in a cloud of concrete dust.

The joyful crowd of thousands surged through the gap to meet their countrymen in East Berlin, which had been sealed off since 1961. Flares and bottle rockets exploded with a candy-pink glare. Young people sprayed the crowd with champagne. The once feared border guards of the Communist DDR stood back, bewildered at the spectacle. They had no orders to intervene, and without orders, they were immobilized, unable to act.

The scene jumped to Checkpoint Charlie, once the espionage fault line between the Soviet bloc and the West. A cavalcade of smoke-belching Trabant minicars trundled cautiously forward from East Berlin, their occupants staring with wonder at the television lights.

I was viewing the spectacle with amazement when a secretary discreetly entered my outer office and whispered, "You're wanted over in SE Division."

When I punched myself through the Division suite's keypad, I found a handful of officers passing stashed bottles of bourbon and scotch around for a modest toast to this long-hoped-for, but largely unanticipated triumph. One by one, people converged from the CIA's two massive Headquarters buildings. Someone cracked open a box of contraband Cuban cigars, and pungent smoke filled the air, in spite of the newly imposed federal ban on smoking in government offices. This afternoon, after all, was a time when exceptions could be made.

As the scene in Berlin became even more frenzied, almost two hundred guests crowded the office suite, men and women who had served on the Agency's front lines. We represented only a tiny portion of the CIA and intelligence community who had helped fight, and win, the Cold War. But among our exclusived up were many who had served with brave, unsung agents like TRINITY who had not survived.

Opinions on the roots of our victory flowed freely. "We knew they were out of fuel for their tanks," a case officer specializing in balance of forces stated confidently. "They just ran out of money."

The political action and propaganda experts had another opinion. "Once we understood how to export the truth about the West directly to the people, the old system couldn't be maintained."

One of my Magic Kingdom wizards spoke up. "The proliferation of information technology made the Iron Curtain obsolete."

"Nah," a strategic conspiracy buff offered. "It was SDI that broke their back. Star Wars was the greatest damn deception in history. We suckered them into spending billions on missile defense when we knew it would never work. Reagan was a helluva poker player."

I moved around the rooms, greeting old friends I hadn't seen for years. Slowly, the realization that the longest, most dangerous espionage confrontation in history was coming to an end dawned on all of us.

But not everyone was convinced, including an indignant counterintelligence officer. "I don't trust the bastards. They've still got twenty thousand nukes and their finger on the button. And the KGB still outnumbers us four to one."

He had a point. No matter what the future of the Soviet Union, the KGB's vast bureaucracy had produced an elite cadre of clandestine operators who might one day sell their services to the highest bidder. His lament, however, was the exception that afternoon. We had won the Cold War, of that I was certain.

But where did this victory take us? With the crumbling of the Soviet bloc, we had lost an evil foe which we had considered imprognable. This unexpected seismic disruption would leave the world unstable and vulnerable to dangerous aftershocks. The ability of the CIA to assess and quickly adapt to this new environment would be critical. There was no time now for a ticker-tape parade, and as the party finally broke up, we had to face the reality of returning to work the next morning and continuing to do honor to our profession—the collection of intelligence in the service of democracy.

CIA Headquarters, November 30, 1990 ■ I pushed the dolly stacked with boxes of my personal effects across the wet leaves of the parking lot. My retirement ceremony had been brief but pleasant, and I was now ready to face the future, knowing that my twenty-five-year CIA career had coincided with the Agency's greatest triumphs.

I deeply regretted that Karen, who had borne so much of the burden of those early years, was not here. But I was happy and confident that Jonna and I, now having decided to share our lives, faced a world much less threatening than the one gripped by the Cold War, which had cast a shadow over the previous five decades.

Another officer surprised me by pushing his own dolly toward his car. The man was an old Soviet–East European Division warhorse who had survived multiple assignments behind the Iron Curtain. He, too, had retired that day.

"What are your plans?" I asked.

"Marge and I have tickets on the Trans-Siberian Express from Moscow to Vladivostok," he said happily. "We're going as tourists."

I waited politely for the punch line. There was none. The man was serious. He and his wife would travel through the vast heart of Russia, so long hidden to Western eyes.

For some reason, I was unexpectedly moved.

■ Epilogue

IN DECEMBER 1990, A FEW WEEKS AFTER MY RETIREMENT, I HAD flown to Germany to join Jonna in Berlin. She was on temporary duty in Moscow, so this quintessential espionage Ground Zero of the Cold War seemed an ideal place for us to meet.

But I experienced a moment of panic when I received word that her flight was diverted from the airport in the former West Berlin to Schöne-feld in the former East Germany. *Hell,* I thought, *who screwed up by sending her into the DDR?* But then I relaxed. The Deutsche Demokra-tische Republik no longer existed. Germany had been reunited under a truly democratic form of government. The former Soviet client states of Eastern Europe were rapidly breaking free of Moscow's imperial control.

Very early on Christmas Eve morning, we walked from our hotel near the Ku'damm all the way to the Brandenburg Gate, where the stark tyranny of the Wall had first been breached and the death of Commu-nism had become inevitable. Even this early on a freezing winter day, the hungry soldiers of the Soviet Army had set up their flea market in the Potsdamer Platz in front of the war-scarred arches. They were ped-dling hammer-and-sickle badges, regimental insignia, and their gleam-ing Red Army belt buckles for a few marks so they could afford a

German breakfast, a piece of wurst and a few potatoes for supper. We bought several belts and a couple of insignia patches, then tried on their steel-gray fur hats and took pictures of each other.

Next we walked through the gap in the wall and down the east side, examining the graffiti and primitive utopian art plastered on the concrete all the way to Checkpoint Charlie. We immediately noticed the inescapable drabness of East Berlin, which stood in gaunt contrast to the holiday prosperity of the West, even a full year after the Wall had been opened.

The people had the same furtive, downtrodden expressions I had first encountered in Moscow in the mid-seventies. Communism had not simply deprived them of material wealth, but it had also crushed their spirits. Most of the men and women with whom we mingled on the way to Unter den Linden had never known a life free of the arbitrary cruelty of the Gestapo or the Stasi. The German Federal government had just announced plans to demolish all traces of the hated Wall. I hoped that they would leave at least one section intact as a bleak memorial to all those people who had suffered under totalitarian Communism.

Jonna and I were married in July 1991. Jesse Lee, her first child, and my fourth, was born on September 19, 1992, and Jonna retired from the CIA four months later.

We returned to Berlin in the summer of 1998 and found the changes in the former East Berlin astounding. We counted more than 150 hammerhead construction cranes radiating out from the Potsdamer Platz, marking the biggest construction boom in European history. Berlin will be reinstated as the capital of Germany in the year 2000. Fortunately, the local government did decide to preserve one small segment of the Wall, not far from Checkpoint Charlie.

Given the undeniable prosperity of Berlin, it would seem that the world has become a safer place now that the West has won the Cold War, and the CIA and its sister services have defeated the KGB and its many allies.

But such optimism may be unfounded.

The 100,000 people who once worked for the KGB did not simply retire to their dachas and take up gardening. True, many are forced to maintain vegetable plots, but their motive is survival. The SVR, or Russian Foreign Intelligence Service, still spies on the West. Alumni of the KGB have blatantly sold themselves, both at home and abroad. Reportedly, members of the elite Grupa Alpha antiterrorism unit of the former KGB Seventh Chief Directorate are now the bodyguards and enforcers of Moscow's leading Mafia bosses. And, of course, Russia still possesses one of the world's largest arsenals of nuclear, chemical, and biological weapons.

I cite these harsh realities whenever people ask if there is still a valid reason for the United States to maintain an extensive, active, and expensive intelligence community.

Yes. Vigilance is, indeed, the price of liberty.

Glossary

Abwehr—The German intelligence service from the 1920s until 1944.

Action officer—The case officer designated to perform an operational act during a clandestine operation especially in a hostile area.

Agent—A person, usually a foreign national, who has been recruited by a staff officer from an intelligence service to perform clandestine missions.

Agent-in-place—An agent serving as a penetration into an intelligence target.

Air America—A CIA proprietary company that provided air support during the secret war in Laos. It has since been sold.

Artist/Validator—An artist trained in forgery working for the CIA's Technical Service.

ARVN—The army of the Republic of South Vietnam.

Asset—A clandestine source or method, usually an agent.

Audio surveillance operation—A clandestine eavesdropping procedure, usually with electronic devices.

Bang and burn—Demolition and sabotage operations.

Black operations—Clandestine or covert operations not attributable to the organization carrying them out.

BND (Bundesnachrichtendienst)—The West German foreign intelligence service established in 1956.

Bona fides—An operative's true identity, affiliation, or intentions. His authenticity.

Bridge agent—An agent who acts as a courier or go-between from a case officer to an agent in a denied area.

Brush pass—A brief encounter where something is passed between case officer and agent.

Burned—When a case officer or agent is compromised.

Cachet—An official mark or wet seal impression made on a document, usually with a rubber stamp or carved seal.

Camp Swampy—A euphemism for the CIA's secret domestic training base, also known as the Farm.

Case officer—A staff officer who runs operations.

Center (Moscow)—KGB headquarters in Moscow.

CHAOS—An ill-advised surveillance operation run by the FBI and the CIA. U.S. antiwar movements during the Vietnam era were suspected of being Communist-inspired.

Cheka—Russian secret police founded in 1917 to serve the Bolshevik party; one of the many forerunners of the KGB.

Chief of Station (COS)—The officer in charge at a CIA station, usually in a foreign capital.

Chieu Hoi—The "Open Arms" rallier program in Indochina used to encourage Communist troops to defect.

Chops—Cachets carved in wood or soapstone.

Cipher—A code wherein numbers or letters are systematically substituted for open text.

Clandestine operation—An intelligence operation designed to remain secret as long as possible.

Clandestine Service(s)—The operational arm of the CIA responsible for classic espionage operations, usually with human assets. Also known as the Directorate of Operations (DO), formerly the Directorate of Plans (DP).

CLOAK—A sensitive disguise and deception illusionary technique first deployed by the CIA in Moscow during the mid 1970s.

Code—A system used to obscure a message by use of a cipher or by using a mark, symbol, sound, or any innocuous verse or piece of music. (Two lanterns in the church tower . . .)

COMINT—Communications intelligence, usually gathered by technical interception and code breaking but also by use of human agents and surreptitious entry.

Commo plan—The various secret communications methods employed with a particular agent.

Compromised—When an operation, asset, or agent is uncovered and cannot remain secret.

Concealment device—Any one of a variety of devices used to secretly store and transport materials relating to an operation.

Controller—Often used interchangeably with handler but usually means a hostile force is involved, i.e., the agent has come under control of the opposition.

Covert Action (CA) operation—An operation kept secret for a finite period of time. Also when the real source remains secret because the operation is attributed to another service.

Cut-out—A mechanism or person used to create a compartment between the members of an operation but allow them to pass material or messages securely.

DAGGER—A sophisticated disguise first used in the Soviet Union in the 1970s.

DCI—The Director of Central Intelligence.

DDO—The Deputy Director of Operations of the CIA, the head of all HUMINT operations, formerly the DDP.

DDP—The Deputy Director of Plans (see *DDO*).

DDR (Deutsche Demokratische Republik)—Communist East Germany.

Dead drop—A secret location where materials can be left in a "concealment" for another party to retrieve. This eliminates the need for personal contact in hostile situations.

Dead telephone—A signal or code passed with the telephone without speaking.

Defector—A person of intelligence value who volunteers to work for another intelligence service. The person may be requesting asylum or can remain in place.

DGI (Dirección General de Inteligencia)—Cuban intelligence service.

DIRECTOR—The cable address of CIA Headquarters.

DIRTECH—The Headquarters cable address of the Office of Technical Service.

DMZ—Demilitarized zone between North and South Vietnam.

Double agent—An agent who has come under the control of another intelligence service and is being used against his original handlers.

Dzerzhinsky Square—Historic site of Lubyanka Prison in Moscow, longtime headquarters of the Soviet security organs including the NKVD and KGB.

EEI (Essential Elements of Information)—An outline to be used for collecting intelligence on a particular topic.

ELINT—Electronic intelligence usually collected by technical interception such as telemetry from a rocket launch collected at a distance.

Escort—The operations officer assigned to lead a defector along an exfiltration route.

EXCOM—The Executive Committee of the CIA made up of the deputy directors and chaired by the Executive Director.

Exfiltration operation—A clandestine rescue operation designed to bring a defector, refugee, or an operative and his or her family out of harm's way.

Expats—Expatriates taking up residence in another land and helping to define its culture.

Family Jewels—The list of "Questionable Activities" that came to light during the Senate investigations of the CIA during the 1970s.

FINESSE—Sensitive disguise developed by the CIA using a Hollywood consultant and contractors.

First Chief Directorate—The foreign intelligence arm of the KGB.

Flaps and Seals—The tradecraft involved when making surreptitious openings and closings of envelopes, seals, and secure pouches.

FLASH—The highest precedence for CIA cable communications.

Foots—Members of a surveillance team who are working on foot.

FSB—Internal security service in Russia, successor to the KGB's Second Chief Directorate (counterintelligence).

GAMBIT—A disguise technique first deployed by the CIA in Southeast Asia in the early 1970s with the help of a Hollywood consultant. It made a quick ethnic change possible.

GRU—Soviet Military Intelligence Service.

Handler—A case officer who is responsible for handling an agent in an operation.

Hmong (the People)—Members of the hill tribes in Southeast Asia.

Ho Chi Minh Trail—Route used by the Communist forces to more or less securely move men and matériel from North to South Vietnam during the war.

Hostile (service, surveillance, etc.)—Term used to describe the organizations and activities of the opposition services.

HUMINT—Human intelligence, collected by human sources such as agents.

ICBM (intercontinental ballistic missile)—Long-range rocket, usually highly accurate and carrying multiple nuclear warheads.

Illegal—A KGB or SVR operative infiltrated into a target country and operating without the protection of diplomatic immunity.

IMMEDIATE—The second highest precedence for CIA cable communications.

Impersonal communications—Secret communication techniques used between a case officer and an agent when no physical contact is possible or desired.

Infiltration—Secretly or covertly moving an operative into a target area with the intent that their presence will go undetected for an appropriate amount of time.

Intelligence Star for Valor—The second highest award for valor given by the CIA.

Internal Operations—CIA operations inside the Soviet bloc during the Cold War.

IO (International Operations) Course—A special training course developed for officers being assigned to the Soviet bloc.

JEDBURGHS—OSS and SOE teams dropped into Fortress Europe after D-Day to help resistance organizations.

KGB—All-powerful intelligence and security service of the USSR during the Cold War. Ultimate successor of Cheka. Disbanded into the SVR and the FSB in 1991.

Kometeh—The Committee of the Islamic Revolution in Iran under Khomeini.

Legend—The complete cover story developed for an operative.

Lima sites—Landing sites built by the CIA on mountain tops in Laos during the secret war. These were the bases of operations behind enemy lines.

"L"-Pill—A lethal cyanide capsule issued to intelligence operatives who would prefer to take their own life rather than be caught and tortured.

MACV—The U.S. Military Assistance Command, Vietnam.

Major docs—The principal identity documents used to authenticate an alias identity.

MI5—The British domestic counterintelligence service.

MI6—The British foreign intelligence service.

Microdot—A photographic reduction of a secret message so small it can be hidden in plain sight or buried under the period at the end of this sentence.

Mili-man—A militia man, a member of the national police force under the Soviet Ministry of Internal Affairs.

Mole—A human penetration into an intelligence service or other highly sensitive organization. Quite often a mole is a defector who agrees to work in place.

Moscow rules—The ultimate tradecraft methods developed for use in the most hostile operational environments. During the Cold War, Moscow was considered the most difficult of the operating environments.

Mossad—Israel's foreign intelligence service.

NE Division—The Near East Division of the CIA's Directorate of Operations.

NIACT—The CIA cable slug that indicates "night action" is necessary.

NKVD—Soviet security and intelligence organization 1934–46.

NOC—A CIA case officer operating under nonofficial cover similar to the KGB illegal.

NVA—North Vietnamese Army.

OGPU—Soviet intelligence and security organization from 1923 until 1934.

Okhrana—Secret police under the Russian tsars 1881–1917.

One-time pad (OTP)—Sheets of paper or silk printed with random five-number group ciphers to be used to encode and decode enciphered messages.

OPS Course (Operations Course)—The elite eighteen-week course all CIA case officers take at the beginning of their careers.

OPS FAM Course (Operations Familiarization Course)—A six-week course for CIA staffers who work with case officers in the field.

OSS—The Office of Strategic Service; forerunner of the CIA, 1942–45.

OTS—The Office of Technical Service, formerly the Technical Services Division, the CIA's technical arm for the Clandestine Service. Develops and deploys technical tradecraft needed for clandestine and covert operations.

OWVL—One-way voice link; shortwave radio link used to transmit prerecorded enciphered messages to an operative.

Pathet Lao—The Communist forces in Laos that joined with the invading North Vietnamese Army during the Vietnam War.

Pattern—The behavior and daily routine of an operative that makes his identity unique.

PHOTINT—Photographic intelligence; renamed IMINT, image intelligence. Usually involving high-altitude reconnaissance using spy satellites or aircraft.

Pipeliner—A case officer or other staff officer who has been designated for an assignment in the Soviet bloc and is undergoing specialized tradecraft training.

Prober—An operative assigned to test border controls before an exfiltration is mounted.

Provocateur—An operative sent to incite a target group to action for purposes of entrapping or embarrassing them.

Q Branch—The fictional part of MI6 that provided spy gadgetry to James Bond. OTS is the CIA's true life "Q."

Questionable Activities—The list of possibly illegal or embarrassing activities compiled throughout CIA during the Senate investigations. These were delivered to Congress by DCI Colby.

Repro—Making a false document.

Resident—The KGB chief of station in any foreign location.

Residentura—The KGB station usually located in the Soviet embassy in a foreign capital.

Revolutionary Guards—The paramilitary security forces serving the Kometeh in Iran.

Road watch teams—Lao irregulars led by CIA officers watching and reporting the enemy's movements.

Rolled up—When the operation goes bad and the agent is arrested.

Roll-out—A surreptitious technique of rolling out the contents of a letter without opening it. It can be done with two knitting needles or a split chopstick.

Safe house—A building or apartment considered safe for use by operatives as a base of operations or for a personal meeting.

SB—Special Branch; usually the national internal security and counterintelligence service.

SDR—Surveillance detection run; a route designed to erode or flush out surveillance without alerting them to the operative's purpose.

Second Chief Directorate—The counterintelligence arm of the KGB responsible for domestic security.

Security service—Usually an internal counterintelligence service, but some have foreign intelligence gathering responsibility such as the Stasi.

SE Division—The Soviet–East European Division of the CIA's Directorate of Operations during the Cold War.

Seventh Chief Directorate—The internal surveillance arm of the KGB. These are the watchers that include the mobile surveillance teams and the technical eavesdroppers.

SIGINT—Signals intelligence; the amalgamation of COMINT and ELINT into one unit of intelligence gathering dealing with all electronic data transmissions.

Silver Bullet—The special disguise and deception tradecraft techniques developed under Moscow rules to help the CIA penetrate the KGB's security perimeter in Moscow.

SITREP—Situation report sent in to Headquarters during an operation or crisis.

SOG (Special Operations Group)—The CIA/SOG was in the Directorate of Operations. The Special Operations Group (c. 1964) of the Department of Defense had a similar paramilitary mission, but there was no connection between the two.

Staff agent—A CIA staff officer without access to CIA secure facilities or classified communications.

Stage management—A vital component to a deception operation: managing the operational stage so all conditions and contingencies are considered—the point of view of the hostile forces, casual observers, physical and cultural environment, etc.

Stasi—East German State Security, including internal security and foreign intelligence.

Station—A CIA office for field operations, usually in a foreign location.

Striker teams—Teams of Lao irregulars led by CIA case officers whose mission was hit-and-run operations against the Communist forces in Laos.

Surreptitious Entry Unit—Unit in the CIA's Technical Service Division whose specialty was opening locks and gaining access to secure facilities in support of audio operations.

SVR—Russian foreign intelligence service succeeding the KGB's First Chief Directorate.

TDY—Temporary Duty Assignment.

Techs—A term used to refer to the technical officers from OTS.

Timed drop—A dead drop that will be retrieved if not picked up by the recipient after a set time period.

Tosses (hand, vehicular)—Tradecraft techniques for emplacing drops by tossing them while on the move.

Tradecraft—The methods developed by intelligence operatives to conduct their operations.

TSD—(see *OTS*)

Walk-in—A defector who declares his intentions by walking into an official installation and asking for political asylum or volunteering to work in place.

Window dressing—Ancillary materials that are included in a cover story or deception operation to help convince the opposition or other casual observers that what they are observing is genuine.